GUIDE [illegible] NG THE NEXT GENE[illegible] STANDARDS

Committee on Guidance on Implementing the Next Generation Science Standards

Board on Science Education

Division of Behavioral and Social Sciences and Education

NATIONAL RESEARCH COUNCIL
OF THE NATIONAL ACADEMIES

THE NATIONAL ACADEMIES PRESS
Washington, D.C.
www.nap.edu

THE NATIONAL ACADEMIES PRESS **500 Fifth Street, NW** **Washington, DC 20001**

NOTICE: The project that is the subject of this report was approved by the Governing Board of the National Research Council, whose members are drawn from the councils of the National Academy of Sciences, the National Academy of Engineering, and the Institute of Medicine. The members of the committee responsible for the report were chosen for their special competences and with regard for appropriate balance.

This study was supported by Contract/Grant No. DRL-1321864 between the National Academy of Sciences and the National Science Foundation, Contract/Grant No. 1012461 from the Burroughs Wellcome Foundation, and a grant from The College Board. Any opinions, findings, conclusions, or recommendations expressed in this publication are those of the author(s) and do not necessarily reflect the views of the organizations or agencies that provided support for the project.

International Standard Book Number-13: 978-0-309-30512-9
International Standard Book Number-10: 0-309-30512-8
Library of Congress Control Number: 2015932191

Additional copies of this report are available from the National Academies Press, 500 Fifth Street, NW, Keck 360, Washington, DC 20001; (800) 624-6242 or (202) 334-3313; http://www.nap.edu.

Printed in the United States of America

Suggested citation: National Research Council. (2015). *Guide to Implementing the Next Generation Science Standards.* Committee on Guidance on Implementing the Next Generation Science Standards. Board on Science Education, Division of Behavioral and Social Sciences and Education, Washington, DC: The National Academies Press.

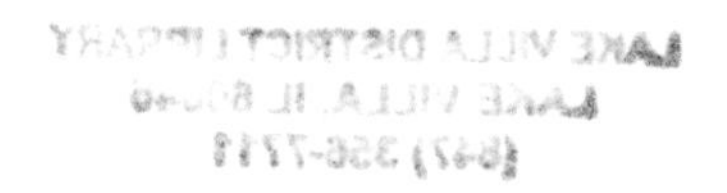

THE NATIONAL ACADEMIES

Advisers to the Nation on Science, Engineering, and Medicine

The **National Academy of Sciences** is a private, nonprofit, self-perpetuating society of distinguished scholars engaged in scientific and engineering research, dedicated to the furtherance of science and technology and to their use for the general welfare. Upon the authority of the charter granted to it by the Congress in 1863, the Academy has a mandate that requires it to advise the federal government on scientific and technical matters. Dr. Ralph J. Cicerone is president of the National Academy of Sciences.

The **National Academy of Engineering** was established in 1964, under the charter of the National Academy of Sciences, as a parallel organization of outstanding engineers. It is autonomous in its administration and in the selection of its members, sharing with the National Academy of Sciences the responsibility for advising the federal government. The National Academy of Engineering also sponsors engineering programs aimed at meeting national needs, encourages education and research, and recognizes the superior achievements of engineers. Dr. C. D. Mote, Jr., is president of the National Academy of Engineering.

The **Institute of Medicine** was established in 1970 by the National Academy of Sciences to secure the services of eminent members of appropriate professions in the examination of policy matters pertaining to the health of the public. The Institute acts under the responsibility given to the National Academy of Sciences by its congressional charter to be an adviser to the federal government and, upon its own initiative, to identify issues of medical care, research, and education. Dr. Victor J. Dzau is president of the Institute of Medicine.

The **National Research Council** was organized by the National Academy of Sciences in 1916 to associate the broad community of science and technology with the Academy's purposes of furthering knowledge and advising the federal government. Functioning in accordance with general policies determined by the Academy, the Council has become the principal operating agency of both the National Academy of Sciences and the National Academy of Engineering in providing services to the government, the public, and the scientific and engineering communities. The Council is administered jointly by both Academies and the Institute of Medicine. Dr. Ralph J. Cicerone and Dr. C. D. Mote, Jr., are chair and vice chair, respectively, of the National Research Council.

www.national-academies.org

COMMITTEE ON GUIDANCE ON IMPLEMENTING THE NEXT GENERATION SCIENCE STANDARDS

HELEN QUINN (*Chair*), Emerita, SLAC National Accelerator Laboratory, Stanford University
MATTHEW KREHBIEL, Kansas State Department of Education
MICHAEL LACH, Center for Elementary Mathematics and Science Education and Urban Education Institute, University of Chicago
BRIAN J. REISER, School of Education and Social Policy, Northwestern University
MARSHALL S. SMITH, Carnegie Foundation for the Advancement of Teaching, Stanford, CA
CARY SNEIDER, Center for Science Education, Portland State University, OR
ROBERTA TANNER, Retired Physics Teacher, Thompson School District, Loveland, CO

HEIDI SCHWEINGRUBER, *Study Director*
REBECCA KRONE, *Program Associate* (until April 2014)
JOANNA ROBERTS, *Program Assistant* (since April 2014)

BOARD ON SCIENCE EDUCATION

Acknowledgment of Reviewers

This report has been reviewed in draft form by individuals chosen for their diverse perspectives and technical expertise, in accordance with procedures approved by the National Research Council (NRC). The purpose of this independent review is to provide candid and critical comments that will assist the institution in making its published report as sound as possible and to ensure that the report meets institutional standards for objectivity, evidence, and responsiveness to the charge. The review comments and draft manuscript remain confidential to protect the integrity of the process.

We thank the following individuals for their review of this report: Juan-Carlos Aguilar, Division of Curriculum, Instruction and Assessment, Georgia Department of Education, Atlanta, Georgia; Philip Bell, Learning Sciences, University of Washington; Rodger W. Bybee, former executive director, Biological Sciences Curriculum Study, Golden, Colorado; Charlene K. Dindo, Baldwin County Education Association, Mobile, Alabama; Ellen Ebert, Teaching and Learning Science, Office of the Superintendent of Public Instruction, Olympia, Washington; George Griffith, superintendent, WaKeeney Unified School District, WaKeeney, Kansas; Jane Hannaway, National Center for Analysis of Longitudinal Data in Education Research, American Institutes for Research, Washington, DC; Kenneth Huff, science teacher, Williamsville Central School District, Mill Middle School, Williamsville, New York; William Penuel, Educational Psychology & Learning Sciences, School of Education, University of Colorado; John Popp, Curriculum and Instruction, Great Bend Unified School District, Great Bend,

Kansas; Carl E. Wieman, Department of Physics and Graduate School of Education, Stanford University.

Although the reviewers listed above provided many constructive comments and suggestions, they were not asked to endorse the content of the report nor did they see the final draft of the report before its release. The review of this report was overseen by Christopher T. Cross, Chairman, Cross & Joftus, LLC, Danville, California, and May R. Berenbaum, Department of Entomology, University of Illinois at Urbana–Champaign. Appointed by the NRC, they were responsible for making certain that an independent examination of this report was carried out in accordance with institutional procedures and that all review comments were carefully considered. Responsibility for the final content of this report rests entirely with the authoring committee and the institution.

CONTENTS

Summary 1

1 **Introduction** 9

The Vision of the *Framework* and the Next Generation Science Standards, 10

Using the Report, 11

Origin of the Report, 12

2 **Overarching Principles for Implementation** 15

Attend to Coherence Across Levels, Across Grades, and Across Different Components of the System, 16

Attend to What Is Unique About Science, 17

Develop and Provide Continuing Support for Leadership in Science at the State, District, and School Levels, 18

Build and Leverage Collaborations, Networks, and Partnerships, 19

Take Enough Time to Implement Well, 20

Make Equity a Priority, 21

Ensure That Communication Is Ongoing and Relevant, 22

3 **Instruction** 23

A Gradual Path, 24

Essential Elements of the Vision of Instruction, 26

Engaging in the Scientific and Engineering Practices, 26

Developing and Using Core Ideas and Crosscutting Concepts, 28
Incorporating Engineering, 29
Creating a Productive Classroom Culture, 30
Connecting Learning Across the Curriculum, 30
Equity, 31
Assessment as an Element of Instruction, 32
Pitfalls to Avoid, 33
Providing Insufficient Support for Students, 33
Focusing Exclusively on the "Right Answers," 33
Assigning Unproductive Student Tasks, 34
Expecting Instruction to Change Overnight, 35
Expecting Teachers to Do It Alone, 35
Being Reluctant to Let Go of Familiar Units or Favorite Activities, 36

4 **Teacher and Leader Learning** 37
Leadership, 38
Learning Opportunities for Teachers, 41
A Focus on Specific Content, 41
Connected to Teachers' Instructional Practice, 43
Active Engagement, 44
Collaboration, 45
Sufficient Time, 45
A Coherent and Ongoing System of Support, 46
Learning Opportunities for Administrators, 46
Pitfalls to Avoid, 47
Underestimating the Shift Needed in One's Own Practice, 47
Underestimating the Need for Ongoing Support, 48
Failing to Provide Opportunities for Administrators to Learn About the NGSS, 49
Offering "One Size Fits All" Learning Opportunities, 49

5 **Curriculum Materials** 51
The Role of Curricula, 52
Developing New Materials, 55
Schools, Districts, and Teachers as Critical Consumers, 56
Pitfalls to Avoid, 58
Asking "Which Standard Are You Teaching Today?", 58

Waiting Before Beginning to Change Instruction, 58
Failing to Provide Resources to Support Students' Investigations and Design Projects, 59

6 **Assessment** **61**
A System of Assessment, 62
Assessment to Support Classroom Instruction, 63
Assessments Designed to Monitor Science Learning on a Broader Scale, 63
Indicators to Track Students' Opportunity to Learn, 64
Implementing a New Assessment System, 64
Pitfalls to Avoid, 67
Failing to Differentiate the Purposes of Assessment, 67
Failing to Respond to Assessment Results, 67
Using Old Assessments While Mandating New Instructional Methods, 67

7 **Collaboration, Networks, and Partnerships** **69**
The Power of Collaboration, 70
Networks for Teachers, 71
Community Partners, 73
Building and Maintaining Partnerships, 76
Pitfalls to Avoid, 77
Lacking a Common Understanding of the Vision, 77
Having Competing Goals Among Partners, 77
Failing to Clarify Relationships and Monitor Partnerships, 77
Failing to Establish Mutually Respectful Relationships and Roles, 78

8 **Policy and Communication** **79**
The Role of Policy, 80
Adequate Time for Learning, 80
Higher Education Admissions, 82
Teacher Preparation and Certification, 82
Communication, 83
Pitfalls to Avoid, 84
Assuming Existing Policies Are Adequate to Support the NGSS, 84
Failing to Communicate with Parents and the Community, 84
Being Unprepared for Unintended Consequences, 84
Assigning Responsibility without Authority or Resources, 85

9 Conclusion 87

References 89

Biographical Sketches of Committee Members and Staff 99

SUMMARY

A Framework for K-12 Science Education: Practices, Crosscutting Concepts, and Core Ideas (National Research Council, 2012; hereafter referred to as "the *Framework*") and the *Next Generation Science Standards: For States, By States* (NGSS Lead States, 2013) describe a new vision for science learning and teaching that is catalyzing improvements in science classrooms across the United States. Achieving this new vision will require time, resources, and ongoing commitment from state, district, and school leaders, as well as classroom teachers.

The Committee on Guidance on Implementing the Next Generation Science Standards of the Board on Science Education was charged with developing guidance for implementing the Next Generation Science Standards (NGSS) as a step toward the goal of ensuring that adoption of the NGSS results in high-quality opportunities to learn science for all students, from kindergarten through high school. The report is intended primarily for district and school leaders and teachers in charge of developing a plan and implementing the NGSS.

PRINCIPLES FOR IMPLEMENTATION

Implementation of the NGSS should be guided by seven principles that reflect the vision of the *Framework*:

1. Ensure coherence across levels (state, district, schools), across grades, and across different components of the system—curriculum, assessment, instruction, and professional development.

2. Attend to what is unique about science.
3. Develop and provide continuing support for leadership in science at the state, district, and school levels.
4. Build and leverage networks, partnerships, and collaborations.
5. Take enough time to implement well.
6. Make equity a priority.
7. Ensure that communication is ongoing and relevant.

RECOMMENDATIONS

To achieve the vision in keeping with the principles, the committee's recommendations cover the major elements in the education system that need to be considered when implementing the NGSS: instruction; professional learning for teachers and district leaders; curriculum resources; assessment; collaboration, networks, and partnerships; and policies and communication. In addition to its recommendations, the committee offers (in the body of the report) pitfalls to avoid for each element.

Instruction

The *Framework* and the NGSS offer a vision of science classrooms where students learn the core ideas and crosscutting concepts of science through engagement in the practices of science and engineering. The nature of instruction required to effectively support the new standards will require changes in many classrooms.

RECOMMENDATION 1 **Communicate and support a vision of instruction that is consistent with *A Framework for K-12 Science Education: Practices, Crosscutting Concepts, and Core Ideas* and the Next Generation Science Standards. Regional and local science education leaders should establish and clearly communicate a vision of science instruction that is consistent with that of the two documents and ensure that their actions, policies, and resource allocations for science education—for professional development, curriculum materials, time to learn, space, equipment, and consumable materials—are aligned to supporting that vision.**

RECOMMENDATION 2 **Support teachers in making incremental and continuing changes to improve instruction. Administrators, science specialists, and resource and mentor teachers should help classroom teachers understand and** adopt the new vision for science learning and instruction through incremental and continuing changes to instruction. They should provide teachers with the curriculum resources needed to support this vision.

RECOMMENDATION 3 Develop a classroom culture that supports the new vision of science education. Teachers should align their teaching approaches, curriculum resources, and students' tasks with the vision. Principals should support the vision and work to provide the necessary resources for teachers and students.

RECOMMENDATION 4 Make assessment part of instruction. Teachers should incorporate performance tasks, open-ended questions, writing tasks, student journals, student discourse, and other formative assessment strategies in their instruction. These activities should be embedded in ongoing classroom work during units and used to obtain information about students' learning in science that can inform further instruction and provide feedback to students. Summative evidence of student learning that is aligned to the performance expectations in the Next Generation Science Standards should be gathered through student work products that document elements of performance tasks.

Teacher and Leader Learning

In many classrooms, instruction will need to change substantially to support the NGSS. In order to understand and support instruction that meets the performance expectations of the NGSS—which integrate scientific and engineering practices, crosscutting concepts, and disciplinary core ideas—both administrators and teachers will need ongoing professional learning opportunities. Teachers will need time and support to transform their instruction.

RECOMMENDATION 5 Begin with leadership. State, district, and school leaders should designate teams that include teachers to lead implementation of the Next Generation Science Standards. Initial professional development efforts should be focused on these leadership teams. Team members should then be engaged in continuing professional learning appropriate to their roles to lead implementation of the necessary changes in curriculum, instruction, and assessment.

RECOMMENDATION 6 Develop comprehensive, multiyear plans to support teachers' and administrators' learning. State, district, and school science education leaders should develop comprehensive multiyear plans for professional learning opportunities for teachers and administrators. These plans need to balance existing resources, meet expectations for milestones in implementation of the Next Generation Science Standards (NGSS), and take advantage of available tools and partners. The plans should take the needs of both current and new teachers into account and allow for ongoing refinement as schools and teachers gain expertise in implementing the NGSS.

RECOMMENDATION 7 Base design of professional development on the best available evidence. When designing professional learning experiences, district and school leaders and providers of professional development should build on the key findings from research. Professional development should (1) be content specific; (2) connect to teacher's own instructional practice; (3) model the instructional approach being learned and ask teachers to analyze examples of it; (4) enable reflective collaboration; and (5) be a sustained element of a comprehensive and continuing support system. For sustained implementation, research shows that principals' understanding of and support for instructional change is key.

RECOMMENDATION 8 Leverage networks and partners. Science education leaders at the state and district level and lead teachers should take full advantage of and cultivate partnerships with other districts, professional development networks, web-based professional development resources, science education researchers, and science-rich institutions—such as higher education institutions and science technology centers—to facilitate high-quality professional development.

Curriculum Materials

Full sequences of curriculum materials designed explicitly for the NGSS have not yet been developed. Until they are available, there are research-based units and materials that support engagement of students in science and engineering practices that can be adapted. In addition, curriculum units currently in use can be revised to be consistent with the NGSS. Teachers will need such materials to support changes in instruction. Individual classroom teachers cannot be expected to develop their own curricula.

RECOMMENDATION 9 Do not rush to completely replace all curriculum materials. States, districts, and schools should not rush to purchase an entirely new set of curriculum materials since many existing materials are not aligned with the Next Generation Science Standards (NGSS). Until new materials are available, district leadership teams in science will need to work with teachers to revise existing units and identify supplemental resources to support the new vision of instruction. In searching for supplemental materials, district leaders and teachers should look for those designed around goals for student learning that are consistent with the NGSS.

RECOMMENDATION 10 Decide on course scope and sequencing. State and district leaders will need to make decisions regarding the scope and sequence of courses in science. Scope and sequence is especially important for grades 6-12, for which the performance expectations of the Next Generation Science Standards (NGSS) are organized in grade bands (6-8 and 9-12). The process of planning scope and sequence should be guided by the strategies outlined in Appendix K of the NGSS.

RECOMMENDATION 11 Be critical consumers of new curriculum materials. District leaders should plan to adopt and invest in curriculum materials developed for the Next Generation Science Standards (NGSS) when high-quality materials become available and in keeping with their own curriculum adoption schedule. District leadership teams should use a clear set of measures and tools with which to judge whether curriculum materials are truly consistent with the goals of *A Framework for K-12 Science Education: Practices, Crosscutting Concepts, and Core Ideas* and the NGSS. Individuals involved in the adoption process should be trained to use those measures and tools.

RECOMMENDATION 12 Attend to coherence in the curriculum. Curriculum designers and curriculum selection teams should ensure that curriculum materials are designed with a coherent trajectory for students' learning. The performance expectations in the Next Generation Science Standards (NGSS) are the target outcomes for the end of a grade level or grade band, and curricula will need to elaborate on a sequence of experiences that will help students meet those expectations. Students need to experience the practices in varied combinations and in multiple contexts to be able to use them as required to meet the NGSS performance expectations.

Assessment

Past assessments of science have chiefly focused on knowledge of facts and procedures, and, hence, are not well suited to the performance expectations of the NGSS. A variety of different assessment and monitoring tools will be needed to serve the different needs of state- and district-level accountability, as well as the needs of classroom-level formative assessment to inform learning and instruction and grading of individual students. All of these will need to be considered in the context of the performance expectations of the NGSS, which integrate scientific and engineering practices, disciplinary core ideas, and crosscutting concepts.

RECOMMENDATION 13 **Create a new system of science assessment and monitoring. State science education leaders should create a long-term plan to develop and implement a new system of state science assessments that are designed to measure the performance expectations in the Next Generation Science Standards. The system should incorporate multiple elements, including on-demand tests, classroom-embedded assessments, and measures of opportunity to learn at the state or district level. When possible, state science education leaders and those responsible for state assessment should consider developing partnerships, perhaps with other states, to facilitate the work of developing new science assessments.**

RECOMMENDATION 14 **Help teachers develop appropriate formative assessment strategies. School leaders need to ensure that professional development for science teachers covers issues of assessment and supports teachers in using formative assessment of student thinking to inform ongoing instruction.**

Collaboration, Networks, and Partnerships

Most states and districts will be facing the same challenges of implementation, and some of the needed expertise resides outside of school systems. Finding, forming, and participating in effective collaborations, networks, and partnerships can facilitate and support the NGSS implementation. The needs at each level of the system (state, district, and school) vary and will require different partnerships and networks. Leaders will need to reach across the traditional boundaries of schools, districts, and states to share information and expertise and identify potential partners, such as informal education institutions, community organizations, and businesses.

RECOMMENDATION 15 Create opportunities for collaboration. District and school leaders should create and systematically support opportunities for teachers and administrators to collaborate within and across districts and schools, with support from relevant experts, with a focus on how to improve instruction to support students' learning as described in *A Framework for K-12 Science Education: Practices, Crosscutting Concepts, and Core Ideas* and the Next Generation Science Standards.

RECOMMENDATION 16 Identify, participate in, and build networks. Science education leaders should identify, participate in, and help build national, regional, or local networks that will enable communities of practitioners, policy makers, science experts, and education researchers to collaboratively solve problems and learn from others' implementation efforts. Teachers and administrators should be encouraged to participate in such networks as appropriate.

RECOMMENDATION 17 Cultivate partnerships. Science education leaders should identify partners in their region and community that have the expertise, motivation or resources to be supportive of their efforts to implement the Next Generation Science Standards and develop relationships with them. In collaboration with potential partners, leaders should determine the kind of support each partner is most suited to provide and to develop strategies for working with them.

Policy and Communication

Policies at the state and district levels and in higher education have complex, interconnected, and often unintended effects. It is important to consider how various policies may affect decisions and opportunities related to implementing the NGSS. Communication and discussion within the education system as well as with external stakeholders at every level is needed to ensure that the goals of the NGSS are understood.

RECOMMENDATION 18 Ensure existing state and local policies are consistent with the goals for implementing the Next Generation Science Standards (NGSS). State boards or commissions with the appropriate authority should review and revise where necessary state-level policies with regard to teacher certification, graduation requirements, and admissions requirements for higher education to ensure they do not create barriers to effective implementation. District leaders should ensure local policies such as teacher assignment support implementation of the NGSS.

RECOMMENDATION 19 Create realistic timelines and monitor progress. State, district, and school leaders should ensure that timelines for implementing the Next Generation Science Standards are realistic and are clearly understood at all levels of the system. They should monitor the implementation and make adjustments when necessary.

RECOMMENDATION 20 Use *A Framework for K-12 Science Education: Practices, Crosscutting Concepts, and Core Ideas* and the Next Generation Science Standards to drive teacher preparation. Provosts, deans, department heads, and faculty in higher education institutions should review and revise programs and requirements for teacher preservice training and introductory undergraduate science courses to ensure these are responsive to teachers' needs under the Next Generation Science Standards, at both the elementary and secondary levels.

RECOMMENDATION 21 Communicate with local stakeholders. State, district, and school leaders should develop a comprehensive strategy for communicating with parents and community members about the Next Generation Science Standards and the changes that will take place to implement them, including a multiyear timeline, possible changes in students' assessment results, and how science classrooms may be different. The communication strategy should include opportunities for public dialogues in which parents and others in the community can provide feedback and express concerns.

1

INTRODUCTION

A Framework for K-12 Science Education: Practices, Crosscutting Concepts, and Core Ideas (National Research Council, 2012; hereafter referred to as "the *Framework*"), and the *Next Generation Science Standards: For States, By States* (NGSS Lead States, 2013) based upon it, have the potential to catalyze improvements in science classrooms across the United States. Together, these documents present a vision of science and engineering learning designed to bring these subjects alive for all students, emphasizing the satisfaction of pursuing compelling questions and the joy of discovery and invention. Achieving this vision in all science classrooms will be a major undertaking and will require changes to many aspects of science education. Effective implementation requires coordinated planning, roll-out of changes across multiple levels of the education system, and sustained efforts to understand and improve practice. This report identifies many of the major challenges and offers guidance to states, districts, and schools on how to plan and implement needed changes.

Standards alone accomplish very little. But standards can help drive improvements when they inform all aspects of the education system, including curriculum scope and sequence, curriculum resources, instruction, assessments, professional development for teachers and administrators, and state policies (National Research Council, 2002, 2006b, 2012). Coordinating changes in all these aspects of the education system is challenging. Improved learning experiences for all students in all classrooms will not occur unless states, districts, and schools develop and follow plans for implementation that allow sufficient time and provide sufficient support to make the necessary changes in a systematic and iterative approach. Such plans will need to be sensitive to and coordinated with other current demands in the system, such as the ongoing efforts in many states to implement new and challenging standards in mathematics and English language arts. To increase the capacity of the system to reach this vision of the *Framework* and the Next Generation Science Standards (NGSS), all of the stakeholders in science education will need to work together. The plans will need to involve a wide range of people and institutions, including places that provide informal learning opportunities; scientists and engineers working in business or higher education; science education researchers; and science-rich institutions and organizations, as well as parents and others in the community.

THE VISION OF THE *FRAMEWORK* AND THE NEXT GENERATION SCIENCE STANDARDS

The research on learning science and engineering that informed the *Framework* and the NGSS emphasizes that science and engineering involve both knowing and doing; that developing rich, conceptual understanding is more productive for future learning than simply memorizing discrete facts; and that learning experiences should be designed with coherent progressions over multiple years in mind (National Research Council, 2007). The *Framework* describes broad learning goals for students in terms of three dimensions: scientific and engineering practices, crosscutting concepts, and disciplinary core ideas. It outlines coherent trajectories for students' learning in science that span grades K-12. The *Framework* emphasizes the importance of providing opportunities for *all* students to continually build on and revise their knowledge and abilities through engagement in the practices of science and engineering. The expectation is also that, in doing so, more students and a more diverse group of students will want to continue their education in these areas to become scientists or engineers and, as citizens, will more deeply understand the processes and core ideas of science and engineering (National Research Council, 2007, 2009).

The NGSS describe ambitious targets for student learning in science that are based on the goals described in the *Framework*. These targets are framed as performance expectations that describe how students will use their knowledge as they engage in scientific and engineering practices. To reach these targets, science education will need to change—for educators at all levels as well as for students, and for networks as well as individuals. The necessary transformations in classrooms will require time, resources, and ongoing attention from state, district, and school leaders.

TABLE 1-1 Implications of the Vision of the *Framework* and the NGSS

Science Education Will Involve Less	Science Education Will Involve More
Rote memorization of facts and terminology	Facts and terminology learned as needed while developing explanations and designing solutions supported by evidence-based arguments and reasoning
Leaning of ideas disconnected from questions about phenomena	Systems thinking and modeling to explain phenomena and to give a context for the ideas to be learned
Teachers providing information to the whole class	Students conducting investigations, solving problems, and engaging in discussions with teachers' guidance
Teachers posing questions with only one right answer	Students discussing open-ended questions that focus on the strength of the evidence used to generate claims
Students reading textbooks and answering questions at the end of the chapter	Students reading multiple sources, including science-related magazines, journal articles, and web-based resources Students developing summaries of information
Preplanned outcomes for "cookbook" laboratories or hands-on activities	Multiple investigations driven by students' questions with a range of possible outcomes that collectively lead to a deep understanding of established core scientific ideas
Worksheets	Students writing journals, reports, posters, media presentations that explain and argue
Oversimplification of activities for students who are perceived to be less able to do science and engineering	Providing supports so that all students can engage in sophisticated science and engineering practices

Together, the two documents provide a vision for science education that both builds on previous national standards for science education and reflects research-based advances in learning and teaching science. This new vision differs in important ways from how science is currently being taught in many classrooms: see Table 1-1.

USING THE REPORT

This report is organized by chapters that correspond to the major elements that need to be considered when implementing the NGSS: Chapters 3-8 cover instruction; professional learning for teachers, administrators, and district leaders; curriculum resources; assessment and accountability; collaborations, networks, and partnerships; and policy and communication (National Research Council, 2002, 2006b). Each of these chapters begins with the committee's recommendations for action and ends with the committee's cautions about potential pitfalls. As context for those chapters, Chapter 2 identifies the overarching principles that should guide the planning and implementation process.

The primary audiences for this report are district and school leaders and teachers charged with developing a plan and implementing the NGSS. Our recommendations are also relevant to a broader audience that includes community stakeholders, out-of-school science program providers, professional development programs, teacher preparation programs, and funders of science education.

Efforts to adopt and then implement the NGSS have been under way since their release in April 2013. Several science education organizations are involved in supporting these efforts including Achieve Inc., the American Association for the Advancement of Science, the Council of State Science Supervisors (especially through its initiative, Building Capacity in State Science Education), the National Research Council's Board on Science Education, and the National Science Teachers Association. Each of these organizations provides online resources that can be helpful in learning about the *Framework* and the NGSS and developing an implementation plan. Websites for these organizations are listed at the end of this report. Some states are already engaged in planning for or implementing the NGSS,[1] but many districts and schools have not yet begun this work.

ORIGIN OF THE REPORT

This report was conceived by the members of the Board on Science Education to help provide guidance for implementing the NGSS over the next decade and beyond. While the NGSS are the focus, the recommendations may also help everyone who is searching for how to better maximize science learning for all students, regardless of their science standards. The seven-member committee was composed of current or past members of the Board on Science Education, which was given the following charge:

> Write a short report regarding necessary steps toward implementation of the Next Generation Science Standards. Drawing on existing National Research Council reports, the report will identify the parts of the education system that need to be attended to when implementing the standards and discuss the changes that need to be made to each part of the system.

[1]See http://ngss.nsta.org/ for updates on which states have adopted the NGSS.

To address this charge, the committee examined the National Research Council reports on science education, as well as those on the broader education system. These sources were supplemented with peer-reviewed research on relevant topics and the members' collective expertise.

2

OVERARCHING PRINCIPLES FOR IMPLEMENTATION

Successful implementation of the Next Generation Science Standards (NGSS) will take a sustained and coordinated effort. It will take multiple years to transition instruction in all classrooms in all schools in a district or state. To be successful, leadership at all levels needs to carefully consider the changes and timeline that will be necessary to move toward the vision for science education laid out in *A Framework for K-12 Science Education: Practices, Crosscutting Concepts, and Core Ideas* (National Research Council, 2012; hereafter referred to as "the *Framework*") on which the NGSS are based.

A first step in planning is to take stock of the current status of each major component of science education activity, both by itself and as part of a whole system, to determine what sequence of decisions and actions is needed and how long each change is likely to take. Some changes, such as starting to involve students in science and engineering practices in science classrooms, can be introduced quite quickly, though they will require more time and attention to be fully developed. Others, such as introducing new statewide assessments that are aligned with the NGSS, will require considerable time for development and testing before implementation (Bybee, 2013; National Research Council, 2014a).

District and school leaders will also need to identify the critical policies and practices that can support or thwart the intended changes and make adjustments to these policies as needed. Examples include a district's adoption or development of particular curriculum materials and allocation of time and resources for teachers' professional development in science. Plans will need to include cultivating support among various communities for any needed policy changes. Those communities include critical actors both within and outside the school system. The key individuals in those communities need to be engaged early and repeatedly in the process, first to plan and later to provide critical feedback and support.

The rest of this chapter discusses seven overarching principles that can help guide planning: coherence across levels and components; the uniqueness of science; continuing support; need for networks, partnerships, and collaborations; sufficient time to implement well; equity; and ongoing and relevant communication. The specific recommendations in the remainder of the report incorporate the principles discussed below. Many of the pitfalls that we discuss in the remaining chapters arise when one or more of the principles are not applied effectively.

ATTEND TO COHERENCE ACROSS LEVELS (STATE, DISTRICT, SCHOOLS), ACROSS GRADES, AND ACROSS DIFFERENT COMPONENTS OF THE SYSTEM (INSTRUCTION, PROFESSIONAL LEARNING, CURRICULUM, AND ASSESSMENT)

Coherence matters (National Research Council, 2006b, 2012). Aligned and coherent supports and an expectation of ongoing collaborative work to understand and implement changes are key to successful reform efforts. The schools and school systems that are improving have all the components working together: tightly interwoven curriculum and assessment are connected to management and evaluation processes, and these in turn drive professional learning at all levels (Smith and O'Day, 1991). Successful implementation of the NGSS requires that all of the components across state, district, and school are aligned to support the vision in the *Framework* and the NGSS.

A standards-based system of science education needs to be coherent in a variety of ways (National Research Council, 2006b, 2012). It needs to be *horizontally coherent:* that is, the curriculum-, instruction-, and assessment-related policies and practices should all be informed by the standards, target the same goals for learning, and work together to support students' development of the knowledge and understanding of science. The system should be *vertically coherent:* that is, there should be a shared understanding at all levels of the system (classroom, school, school district, state) of the goals for science education and agreement about the purposes and uses of assessment.

The system should also be *developmentally coherent:* that is, there needs to be a shared understanding across grade levels of what ideas are important to teach and of how children's understanding of these ideas can develop across grade levels. The *Framework* and the NGSS support developmental coherence by describing how each core idea, practice, and crosscutting concept is expected to develop across the span from kindergarten through high school (K-12). In order to allow students to explore important ideas in science deeply across multiple grades, some topics that are currently taught may receive less emphasis or may need to be eliminated entirely (National Research Council, 2007).

Coherence does not occur accidentally. To achieve it takes planning, political will, professional time, and ongoing management. Leaders need to ensure that those responsible for different components or for different grade levels have the responsibility, opportunity, and authority to work together, rather than each moving ahead in isolation. At each school level or grade level within a school, those responsible for planning and implementing changes need to be aware of what changes are planned and what have already occurred in the earlier grades and also of what will be expected of the students in later grades.

ATTEND TO WHAT IS UNIQUE ABOUT SCIENCE

Implementing science standards is different from implementing standards in English language arts or mathematics, though some challenges will be similar. It is important to build on and coordinate with efforts to implement the new standards in mathematics and English language arts while also attending to how science is different.

Typically, there are fewer individuals with expertise in science and science pedagogy available within the school or district than individuals with comparable expertise in English language arts and mathematics. And many administrators do not have science backgrounds. This kind of expertise is relevant when selecting instructional materials, sequencing curriculum, observing classrooms, and hiring educators. There are also some costs associated with science—for materials or laboratory space—that are different than the costs for mathematics and English language arts. Finally, in many states, science is not as important for school and teacher accountability as the other two subjects and has therefore received less emphasis than they have.

Implementation strategies have to respect and embody the differences between subjects even as they build on their similarities. Some pedagogical and classroom management strategies apply across subjects, while some do not. It is important to consider links between standards in mathematics and English language arts and the NGSS: one is the role of productive student discourse in all three and the changes in classroom culture required to support it (Michaels et al., 2008).

A focus on science may pose particular challenges at the elementary level. In many schools and districts, very little science is currently taught in the elementary grades. According to a national survey of science education conducted by Horizon Research (see Trygstad, 2013), 39 percent of elementary classrooms did not include science every week. Elementary teachers spent, on average, only 20 minutes on science every day. In comparison, they spent 55 minutes for mathematics and 88 minutes for reading.

Furthermore, analysis of 4th-grade data from the 2009 National Assessment of Educational Progress (NAEP) in science showed that time spent on science varies widely by state, ranging from a low of 1.9 hours per week in Oregon to a high of 3.8 hours per week in Kentucky, and that the time spent on science is significantly correlated with achievement in science (Blank, 2013). Data from California showed that 40 percent of elementary teachers spent an hour or less on science per week, and, of those, 13 percent spent less than 30 minutes per week (Dorph et al., 2011).

Ensuring time for science at the elementary level is an important issue and will need to be considered early in the implementation process. That consideration needs to include the possibility of changing policies about time spent exclusively on other subjects, remediation, and the resources needed (such as space and materials) for investigative and design activities. It might also include discussion of how to integrate science, mathematics, and English language arts (see National Academy of Engineering and National Research Council, 2014; National Research Council, 2014b). At the middle and high school levels, laboratory space and materials are more likely to be in place, but their role and use may need to be reconsidered to allow students to engage in the full range of science and engineering practices (National Research Council, 2006a).

DEVELOP AND PROVIDE CONTINUING SUPPORT FOR LEADERSHIP IN SCIENCE AT THE STATE, DISTRICT, AND SCHOOL LEVELS

An early priority is to establish district and school leadership teams that involve a mix of stakeholders (including administrators, teachers, science education researchers, and representatives from the community) who are given the responsibility, resources, authority, and work time needed to lead the implementation effort. And before they can lead and support changes in instruction and curriculum, their learning needs should be addressed, so they can then support the learning needs of all teachers.

Teacher leaders are invaluable for supporting and institutionalizing changes. They work with other teachers and parents, as mentors to other teachers, and as facilitators of reflective learning, in the classroom and in the learning culture of a school (Coburn et al., 2012; Fogleman et al., 2006; Penuel and Riel, 2007; Spillane, 2006a, 2006b; Sun et al., 2013a, 2013b). The NGSS has already generated significant attention in the professional organizations of science teachers, such as the National Science Teachers Association and the National Science Education Leadership Association. Many science teachers are well ahead of their schools and even their states in thinking about the demands on their students that the NGSS will bring and how their own instruction will need to change to prepare their students to meet these demands. Identifying and making use of the teachers who are ready to be the "early adopters," particularly those who may already play leadership roles in the teacher community of a school or district, needs to be a key part of a school's implementation strategy.

BUILD AND LEVERAGE COLLABORATIONS, NETWORKS, AND PARTNERSHIPS

One advantage of a set of standards that will be used across multiple states is the opportunity to share the work through networks, partnerships, and other collaborations across states. Those networks can serve different levels, from the state, to the district, to the school, and individual teachers. Networks of teachers focused on implementing a shared approach to science can be immensely productive—both to the participants and to the teachers they mentor. Collaborations might involve other schools, districts, and states, as well as other stakeholders in the community, such as universities, businesses, museums, and other institutions that can offer resources for science learning. Networks and partnerships can allow schools and districts to access additional science expertise and resources. They also can help build broad community support for the NGSS, including reaching out to the scientific community.

In implementing the NGSS, do not try to go it alone. Changing one classroom in one school will not provide the science learning experiences that all K-12 students should have. Instead, groups of leaders and teachers in different states, districts, and schools, at times in collaboration with businesses or community organizations, can develop strategies and joint resources to help achieve the goals of the NGSS.

TAKE ENOUGH TIME TO IMPLEMENT WELL

Implementing the NGSS will be demanding and will require persistence. The NGSS require that students not only know science facts but can also apply them to explain phenomena or solve problems using the science and engineering practices. In many classrooms, this will represent a significant increase in complexity and cognitive demand for both teachers and students. Achieving such changes will require attention over many years.

Time is needed for the development of appropriate curriculum materials and assessments; for teachers to embrace the expectations of the standards and adapt their instructional strategies to empower students to achieve the level of performance expected; and for students to adjust to new expectations and to build the foundation of knowledge and skills to meet these challenging standards. For example, middle or high school students who enter science classes today are likely to be unfamiliar with many of the science practices in the NGSS, but in 6-8 years students should arrive in middle school with knowledge of those practices and several years of building the skills needed for science at a higher level. Thus, the higher the grade level, the more time it will take before one can imagine that implementation has reached a stable configuration. Even then, ongoing attention will be needed to keep improving science learning for all students.

It may be tempting to expect to see results in students' achievement within 1-2 years, but it will likely take a minimum of 3-4 years for teachers to transition to effectively teaching the new standards. It is essential to allow time for the necessary ingredients, such as professional development, team building, and appropriate curriculum resources, to be in place. Teachers need time and support to develop expertise for teaching in new ways. It takes several years for changes in instruction to become stabilized (Lee et al., 2008; Marx et al., 1998; Supovitz and Turner, 2000, cited in Wilson, 2013). Sustaining such changes depends on a cadre of teachers who are engaged in ongoing reflection on their instructional practice (Coburn et al., 2012; Franke et al., 2001).

It is important for school leaders to be prepared to accept less than perfect outcomes in the initial years of implementation of the NGSS. They will need to prepare teachers and parents to expect the process will take time. It may be helpful to identify interim benchmarks of progress, other than students' achievement, to track progress toward implementation in the classroom. A range of benchmarks might be considered, such as the availability of ongoing professional learning opportunities for teachers that align to the NGSS, the amount of time spent on science in elementary classrooms, the adoption of curriculum materials that provide opportunities to engage in all three dimensions of the NGSS, the numbers of students enrolling in high school science electives, the establishment of a cohesive K-12 district science team with work time structured into the school year, and the development of mutually beneficial relationships with informal science educators or local business or industry (see National Research Council, 2013).

MAKE EQUITY A PRIORITY

The vision of the *Framework* and the NGSS is that *all* students will have access to high-quality learning opportunities in science and will be able to succeed in science (National Research Council, 2012). Thus, one component of implementation will be to track whether changes are supporting equality of opportunity to learn science across all districts in a state, all schools in a district, and all classrooms in a school (National Research Council, 2014a).

An "achievement gap" between students from low-income backgrounds in comparison with students from high-income backgrounds persists in science, as it does in other subjects. For example, on the 2009 NAEP science assessment, 4th-grade students from schools with a high percentage of students on free and reduced lunch scored an average of 28 points lower (on a 300-point scale)—approximately 2.3 grade levels lower—than students from schools with a low percentage of such students. A key to addressing such gaps is to pay continuing attention to issues of opportunity to learn science, with qualified teachers and adequate resources, at all grade levels (National Research Council, 2013, 2014a). Districts' attention to monitoring such opportunities and ensuring access for all students is an important element of implementation planning.

Attention to equity also requires consideration of how to meet the differing needs of students, including those who have special learning needs, do not have access to technology, are learning English as a second language, are living in difficult economic circumstances, or are from nondominant cultural backgrounds. Equity also requires attention to the availability of advanced science courses in high school (through advanced placement, international baccalaureate, or honors courses) for all students who are interested in science and ready to pursue those courses.

The *Framework* provides an indepth discussion of equity, including attention to sources of inequity and providing inclusive science instruction (National Research Council, 2012, Ch. 11). The NGSS discusses equity and diversity in depth, with case studies that offer valuable examples of equitable instruction (NGSS Lead States, 2013, App. D).

ENSURE THAT COMMUNICATION IS ONGOING AND RELEVANT

The best of plans will fail if they are not well communicated to all needed audiences. Districts and schools need to understand their state's timelines and expectations as they develop their own plans. They also need to ensure that their constituents—administrators, teachers, parents, and students—understand and support the implementation process. Maximizing transparency about steps in the plan, when they will happen, and why can help build the support needed to weather the inevitable bumps in any implementation plan. It is also important to emphasize that implementation is a 5-10-year process, and stakeholders need to be supportive of the long-term goals rather than focus solely on short-term results.

Communication about the NGSS and how they will be implemented is only effective when care is taken to ensure that the intended messages are being heard and understood. Words like standards, instruction, curriculum, inquiry, rigorous learning, alignment, and even science have multiple meanings not only among implementation partners in the community and the public, but even in the education community. For example: Is instruction what the teacher says, or is it everything that happens in the classroom? What is a curriculum and what is a curriculum resource?

Stakeholders, working together, will need to come to a common understanding of each of these terms. In this document, we have given some working definitions of what we mean as we use terms such as these. We expect that similar discussions will be needed at the state, district, and school levels to ensure that policies and practices, and internal and external communication about them, are clear and lead to shared understandings.

3

INSTRUCTION

RECOMMENDATION 1 Communicate and support a vision of instruction that is consistent with *A Framework for K-12 Science Education: Practices, Crosscutting Concepts, and Core Ideas* and the Next Generation Science Standards. Regional and local science education leaders should establish and clearly communicate a vision of science instruction that is consistent with that in the two documents ensure that their actions, policies, and resource allocations for science education—for professional development, curriculum materials, time to learn, space, equipment, and consumable materials—are aligned to supporting that vision.

RECOMMENDATION 2 Support teachers in making incremental and continuing changes to improve instruction. Administrators, science specialists, and resource and mentor teachers should help classroom teachers understand and adopt the new vision for science learning and instruction through incremental and continuing changes to instruction. They should provide teachers with the curriculum resources needed to support this vision.

RECOMMENDATION 3 Develop a classroom culture that supports the new vision of science education. Teachers should align their teaching approaches, curriculum resources, and students' tasks with the vision. Principals should support the vision and work to provide the necessary resources for teachers and students.

RECOMMENDATION 4 Make assessment part of instruction. Teachers should incorporate performance tasks, open-ended questions, writing tasks, student journals, student discourse, and other formative assessment strategies in their instruction. These activities should be embedded in ongoing classroom work during units and used to obtain information about students' learning in science that can inform further instruction and provide feedback to students. Summative evidence of student learning aligned to the performance expectations in the Next Generation Science Standards should be gathered through student work products that document elements of performance tasks.

A GRADUAL PATH

A Framework for K-12 Science Education: Practices, Crosscutting Concepts, and Core Ideas (National Research Council, 2012; hereafter referred to as "the *Framework*") and the *Next Generation Science Standards: For States By States* (NGSS Lead States, 2013) do not dictate a single approach to instruction. There are many approaches to science instruction that could be consistent with the vision in those documents. By instruction, we do not mean the information that a teacher delivers to students; rather, we mean the set of activities and experiences that teachers organize in their classroom in order for students to learn what is expected of them. The scope and sequence of these activities should be guided by a curriculum plan and be supported by curriculum resources that are well matched to that plan, while the day-to-day instruction is carried out by teachers who are making continual decisions about what best meets their students' needs along a learning path that allows them to achieve the types of proficiencies and performances embodied in the Next Generation Science Standards (NGSS).

The heart of the *Framework* and the NGSS is a clarification and focusing of what students need to know and to be able to do in science. An important first step for implementation is for both school leaders and teachers to establish a shared vision both of what should be happening in science classrooms to support such learning and of what successful student performance should look like. Only when such a shared vision has been articulated and broadly communicated can the extended effort that will be needed to implement the necessary changes be successful.

The performance expectations in the NGSS are targets for assessment. For students to achieve such performances, they will need regular opportunities to engage in learning that blend all three dimensions of the standards throughout their classroom experiences, from kindergarten through high school (K-12). When instruction is consistent with the *Framework* and the NGSS, students will be actively engaged in the full range of scientific and engineering practices in the context of multiple core ideas. Student work will be driven by questions arising from phenomena or by an engineering design problem. Students will be supported in connecting their learning across units and courses to build a coherent understanding of science ideas and of the crosscutting concepts. They will have opportunities to apply their developing science knowledge to explain phenomena or design solutions to real-world problems. Finally, they will interact with each other as they conduct investigations; represent data; interpret evidence; gather additional information; and develop explanations, models, and arguments.

Many teachers will need time and support to transform their instruction so that it reflects this vision (Banilower et al., 2007; Reiser, 2013). That support should include, but not be limited to, ongoing professional learning opportunities for both teachers and administrators to create a shared understanding of goals for instruction and to collaborate on steps to achieve them. Teachers and district science leaders will need to work together to reevaluate the scope and sequence of the science that they teach, their curriculum materials, unit and lesson plans, and the classroom-level assessment tasks that they use to make sure that these are all designed to support the multidimensional learning outcomes expected for the NGSS. Teachers need structured time to engage with others in ongoing evaluation of the effectiveness of their approaches for helping students achieve the instructional goals. They also need structured time to reconsider and revise those goals, and they need district policies that are supportive of the changes they are expected to make (see Chapter 4 for a detailed discussion of teacher learning).

It is unrealistic to expect teachers to completely transform their instruction at one time or quickly. They will need time and ongoing support to take incremental steps toward the instructional vision, over a period of at least 2-3 years. For example, teachers might start by teaching only one new or redesigned unit that incorporates science and engineering practices and focuses more in depth on the target disciplinary core idea. Even after the initial 2-3-year implementation period, continued support for teachers will be important, through participation in a professional learning community, for example, as teachers refine their instructional approaches. When new curriculum materials are developed, adopted, or purchased, teachers will need time for professional development and collaboration in order to use the new resources effectively. Classroom and school budgets will need to support the purchase of the equipment and supplies that are required to implement the new curriculum.

ESSENTIAL ELEMENTS OF THE VISION OF INSTRUCTION

This section provides a sketch of what instruction developed to support the NGSS might look like and describes some of the changes that will be required in classrooms (see Chapter 9 of the *Framework,* National Research Council, 2012, for an additional discussion).

Engaging in the Scientific and Engineering Practices

The science and engineering practices in the *Framework* and the NGSS elaborate on what it means to engage in scientific inquiry and engineering design. Engaging in these practices helps students understand how scientific knowledge develops and gives them an appreciation of the wide range of approaches that are used by scientists to investigate, model, and explain phenomena in the natural world and in engineered systems (National Research Council, 2012). The science and engineering practices also help students develop capabilities in engineering design, which includes defining and solving problems. Furthermore, students' engagement in these practices is a critical element of supporting the conceptual changes (that is, changes in students' ideas about the world) that are required for students to develop and deepen their understanding of the core ideas and crosscutting concepts of science (for an indepth discussion of conceptual change, see National Research Council, 2007, pp. 106-120).

Students need to have multiple opportunities to ask questions about, investigate, and seek to explain phenomena, as well as to apply their understanding to engineering problems. Students' ideas are learned more deeply and retained longer if students apply them to situations that have meaning for them. In a classroom that is consistent with the *Framework* and the NGSS, students develop models of the phenomenon being studied that make explicit their understanding of both visible and invisible aspects of what is occurring: two examples are interactions at the molecular level that explain the behavior of an air mass in a weather phenomenon and the accumulation of events across time that explain population-level phenomena in ecosystems.

Students apply and improve their understanding of science core ideas and crosscutting concepts as they develop and refine these models. They then use their models and their understanding of the science in question to support their explanations of what occurs or to design solutions for real-world problems. Students analyze evidence and engage in model- and evidence-based argumentation to support or critique an explanation, respond to critique of their own ideas, and compare the merits of alternate design solutions.

Importantly, the scientific and engineering practices work in concert with each other; they are not intended to be learned in isolation from each other. For example, as students analyze data they will likely use some mathematics. As they generate, discuss, and critique explanations, they will rely on model-based and evidence-based argumentation and reasoning. As they design and carry out investigations, they will need to revisit and refine their initial questions. And as they obtain and evaluate information from multiple sources, they will need to ask questions about what they are reading and its sources. The practices are neither a set of steps in a process nor a recipe as to how to proceed; rather, they are tools to be used as needed, and often one needs more than one tool at a time for a question or problem. It is also important to emphasize that a student's ability to memorize facts, formulas, and definitions should not be a prior condition for engaging in the practices; rather, it is through developing models and explanations and engaging in argumentation to refine and improve explanations that students come to understand the value and meaning of definitions and facts (National Research Council, 2000).

The purpose of having students engage in the science and engineering practices around real-world phenomena is not that students will discover the science ideas for themselves. The phenomena or design problems introduced have to be carefully chosen to provide a context in which students become engaged and in which the science ideas they are learning are useful because they can help explain what is occurring. Students still need to learn basic science ideas and terminology, whether through reading about them or through a teacher's questions, suggestions, and focused explanations, in order to be able to use them within their models and in developing their explanations about the phenomena. Students who learn and apply science ideas in this way integrate the ideas more deeply into their view of the world, are more likely to apply them for problem solving in new contexts, and remember the ideas longer than those who simply learn them as "facts" discovered by scientists that need to be memorized (National Research Council, 2000, 2007).

The *Framework* and the NGSS are also explicit about the need to engage students in using a range of technologies, including (but not exclusively) digital technologies. Such tools need to be used purposefully to advance particular learning goals, for example, to help students engage with real data, investigate phenomena, or work with and communicate their models. In particular, the practice of computational thinking involves such activities as simulations to model physical phenomena or test engineering designs under a range of different conditions, to mine existing databases, or to use computer-aided design software to design solutions to problems.

Developing and Using Core Ideas and Crosscutting Concepts

The NGSS are organized around central explanatory ideas in science and engineering for which students develop increasingly sophisticated understandings across K-12. The *Framework* and the NGSS articulate how disciplinary core ideas build coherently across multiple grades and connect between the life, physical, Earth and space sciences, and engineering. For example, students' understanding of matter and its properties develop across the grade levels. In the early grades, these understandings relate chiefly to recognizing and categorizing matter by its properties. Ideas about what changes and what does not as matter interacts or conditions change begin to be developed in these grades and are refined and made more explicit in the subsequent grade levels.

Critical understandings and models of the particle substructure of matter and how this structure changes with conditions, such as undergoing transitions between solids, liquids, and gases, help explain many properties of matter. These understandings are developed in the middle school years, and many aspects of middle and high school science across all disciplines build on these models.

Learning sequences within a grade need to be designed with coherence in mind. This may require a reorganization of topics and omission of some existing units. The goal is to provide students with multiple opportunities to explore important scientific ideas in depth at a level of sophistication appropriate for their grade level. Exploration of the ideas occurs through engagement in the science and engineering practices.

The crosscutting concepts introduced in the *Framework* are relevant across the disciplines of science and can help students make connections across topics, courses, and disciplines. Teachers using appropriately designed curriculum resources can support students in applying the crosscutting concepts across different core ideas and, at the secondary level, across different courses. For example, the idea of a system, and the need to delineate and define a system in order to model it, is used again and again across all of the sciences. By developing a common language and set of questions around this concept, students not only acquire a useful tool for analyzing phenomena or designs, they also develop a view of what is common across very different science disciplines. Using and reflecting on both the science and engineering practices and the crosscutting concepts are thus important elements in developing a deeper understanding of the nature of science and the role of engineering.

Activities need to be sequenced so that students' understanding of core ideas in the disciplines and how they are related through crosscutting concepts develops over one school year and over multiple years.[1] Connecting across grades and across disciplines creates a learning environment in which the significance of ideas for making sense of the world drives learning, rather than external motivators such as "you'll need this next year" or "this will be on the test." The sequence of core ideas that are introduced throughout the year, and the connections made between them, are important in helping students develop an understanding of the most important ideas in science and how they are connected or related through crosscutting concepts.

Incorporating Engineering

In the *Framework* and the NGSS, engineering plays three roles. First, engaging in the engineering practices is a vehicle for building students' understanding of science ideas by applying them to solve engineering problems. For example, designing a toy car to meet a specific performance challenge can provide a context to develop or extend students' understanding of force and motion. These kinds of experiences also help students recognize how science affects their lives and society through engineering and technology. For many students, understanding these

[1]For details, see Chapters 3-8 of *A Framework for K-12 Science Education: Practices, Crosscutting Concepts, and Core Ideas* (National Research Council, 2012) and Appendixes E, F, and G of the *Next Generation Science Standards: For States, By States* (NGSS Lead States, 2013).

effects gives relevance to science and makes science learning a more meaningful pursuit.

Second, engineering design itself is designated as a core idea, defining knowledge that is needed in order to engage in engineering practices. Students are expected to learn a few key engineering concepts, such as the process of design, as they gain facility with using the engineering practices by engaging in engineering design projects. Finally, students will also come to understand the similarities and differences between the ways the practices are used for science and engineering purposes.

Creating a Productive Classroom Culture

In order for students to participate in the full range of science and engineering practices and for them to have time to develop their own explanations, models, and arguments, the structure of classroom activities and discussions will likely need to change. The norms for how students interact with each other and the teacher, how they work on tasks together, and how they respond to each other's ideas will also need to change (Berland and Hammer, 2012; Driver et al., 2000).

Engagement in the science and engineering practices requires social interaction and discussion among students. Students need support to learn how to do this productively. The classroom culture will need to support both individual and collaborative sense-making efforts. Students will learn to take responsibility for their learning rather than waiting for answers, and they will be expected to collaborate with, critique, argue with, and learn from their peers.

A classroom culture that supports this kind of student engagement will likely require a significant shift in classroom management strategies for most teachers (Windschitl et al., 2008). Teachers need to see and analyze examples of how to facilitate discussions that enable all students to participate and learn, and they will need opportunities to try strategies in their own classrooms.

Connecting Learning Across the Curriculum

The combination of the NGSS and the Common Core State Standards in English Language Arts and Mathematics (National Governors Association, 2010) offers opportunities to strengthen students' learning through use of similar strategies across the curriculum. All three sets of standards emphasize student reasoning and arguing from evidence—even though the nature of an effective argument and what counts as evidence is specific to each subject. Science and engineering problems can be used as examples while teaching mathematics. Science topics can be explored through using science-related trade books or magazine articles for reading in language arts classes. These activities can help support science learning, but they cannot provide all of the science learning opportunities that students need. Conversely, engaging in the science practices requires students to apply their mathematics and literacy skills in the context of their science classrooms and so can help students further develop those skills.

While engaging in the scientific and engineering practices, students will regularly construct oral and written arguments that focus on presenting and evaluating evidence for claims, resolving differences, and refining models and explanations or on improving engineering designs. Students will seek and evaluate information from a variety of sources to support and extend their science understandings. They will read, write, and communicate orally about science ideas. Students and teachers will use mathematics and computer-based tools and simulations flexibly and effectively to support investigations, data collections, and analysis and to develop understanding of key concepts.[2]

EQUITY

A Framework for K-12 Science Education and the NGSS emphasize that concerns about equity should be a focus of efforts to improve science education (see National Research Council, 2012, Ch. 11; NGSS Lead States, 2013, Appendix D). All students should have access to high-quality learning experiences in science. The *Framework* focuses on two sources of inequity that can be most directly addressed by educators. The first links differences in achievement to differences in opportunities to learn because of inequities across schools, districts, and communities. The second considers how approaches to instruction can be made more inclusive and motivating for diverse student populations.

Inclusive instructional strategies encompass a range of techniques and approaches that build on students' interests and backgrounds so as to engage students more meaningfully and support them in sustained learning. An important element of many of these approaches is recognizing the assets that students from diverse backgrounds bring to the science classroom and building on them. Such

[2]The NGSS were constructed to facilitate making connections between mathematics and English language arts. In the NGSS, Appendix L discusses connections to mathematics, and Appendix M discusses connections to English language arts. A recent report of the National Research Council (2014b) provides examples of how to connect literacy and science and discusses many of the key issues in making such connections.

assets include students' everyday experiences in their communities, their prior interests, and their cultural knowledge and modes of discourse. Appendix D of the NGSS includes a detailed discussion of these specific instructional approaches with examples that feature students from different groups.

ASSESSMENT AS AN ELEMENT OF INSTRUCTION

Classroom-based assessment activities are critical supports for instruction. Classroom assessments can play an integral role in students' learning experiences and inform subsequent instructional choices, while also providing evidence of progress in that learning. Implementation of the NGSS demands the use of assessment tasks that integrate the dimensions of the *Framework* (National Research Council, 2014a). These tasks also need to be designed so that they can accurately locate students along a sequence of progressively more complex understandings of a core idea and crosscutting concepts and successively more sophisticated engagement in science and engineering practices (National Research Council, 2014a).

Instruction that is consistent with the *Framework* and the NGSS will naturally provide many opportunities for teachers to observe and, on occasion, to record student performances that integrate the dimensions, and to use student work products to reveal student thinking. Science and engineering practices lend themselves well to being used as assessment activities: indeed, the line between instructional activities and assessment activities may often be blurred (National Research Council, 2014a), particularly when the assessment purpose is to inform future instruction rather than to grade individual students (Atkin and Coffey, 2003). Whether assessment opportunities are fully integrated into instruction or are more formal individual assessment tasks, students need guidance about what is expected of them, opportunities to reflect on their performance, and detailed feedback on how to improve their performance. Teachers need to see and work with examples of student work produced in the course of engagement in the science and engineering practices that can be used for assessment purposes. Methods of evaluating students' performance—for example, scoring rubrics—can be developed and used to inform future teaching. Analysis of students' work products and discussion of how to use them for assessment purposes could take place in the context of collaborative lesson study among groups of teachers.

Assessment tasks designed to be seamlessly integrated with classroom instruction are beginning to be developed, some of which are performance tasks. Performance tasks, while forming part of an ongoing learning sequence, also contain elements to be produced by individual students that can be used as summative assessments (e.g., to assign student grades for a unit or course). The early versions of these types of assessments demonstrate that it is possible to design tasks that successfully elicit students' thinking about disciplinary core ideas and crosscutting concepts by engaging them in scientific and engineering practices (National Research Council, 2014a).

Assessments of science learning that integrate practices, crosscutting concepts, and core ideas are challenging to design, implement, and properly interpret. Teachers will need extensive learning opportunities to successfully incorporate both formative and summative assessment tasks that reflect the performance expectations of the NGSS into their practice (National Research Council, 2014a).[3]

PITFALLS TO AVOID

Providing Insufficient Support for Students

As students are asked to learn new practices and engage with science ideas in new ways, they will need "scaffolding"—that is, a set of supports (National Research Council, 2007). It is important to provide sufficient time and support for students to develop increasingly sophisticated explanations of phenomena; to learn to support and explain their arguments with evidence; to make, accept, and respond to their peers' critiques of explanations, models, and designs; and to develop greater facility with all of the practices (Furtak et al., 2012). All of this requires a shift of classroom culture, of pedagogy, and of students' understanding of what it means to learn well. This shift is particularly important as schools, districts, and states think about how to support students in the higher grades who, in early years of implementation, may not have had the prior learning experiences needed to meet the expectations in the NGSS.

Focusing Exclusively on the "Right Answers"

The emphasis in the *Framework* and the NGSS on discussion and allowing time for students to develop arguments and explanations can be uncomfortable for both teachers and students. For teachers, it can be difficult to allow students to explore incorrect or partially correct ideas out of concern that they will never

---

[3]The recent report *Developing Assessments for the Next Generation Science Standards* (National Research Council, 2014a) provides discussion of, and examples of, classroom-based assessments that are consistent with the NGSS. For a detailed discussion of formative and summative assessments, see National Research Council (2001).

arrive at the correct explanation. However, focusing exclusively on right answers can limit students' engagement in argumentation and discourage discussions (Lemke, 1990; Mortimer and Scott, 2003). Teachers need to see how students' models can become more accurate and complete over time, through elaborating, refining, and fine-tuning the models that may at times contain incomplete or technically correct but misleading ideas, rather than seeing students' ideas as simply "correct" or "misconceptions" (Windschitl et al., 2008).

Students who have experienced success in school primarily by memorizing and reproducing facts or rote procedures provided to them by textbooks or teachers may resist the shift to a classroom culture where they are asked to apply science ideas and take part of the responsibility for the struggle to develop shared explanations to make sense of phenomena. Students need to learn about the ideas that have been established over many years of science, but they also should be able to construct evidence-based arguments that support these ideas or that refute alternate and commonly held naïve conceptions. They should be able to apply the scientific ideas in appropriate contexts to explain natural phenomena or design solutions to problems that may have several acceptable solutions.

Both students and teachers run the risk of slipping into the mode of students waiting to be told and of teachers as the purveyors of "right" answers. These tensions should be anticipated and proactively addressed through professional learning for teachers and administrators and well thought out messages shared with the community (including parents and students) before beginning to implement the NGSS. The messages need to be reinforced throughout the process.

Assigning Unproductive Student Tasks

The types of tasks that students are asked to engage in will look different in a classroom aligned to the NGSS. For example, simply memorizing a science vocabulary list—such as the names of parts of a cell or reading a textbook selection and answering questions at the end of the chapter that require students to restate or repeat portions of the text—is not consistent with the vision for learning in the *Framework* and the NGSS. Instead, students could be asked to explain how the function of a particular part of the cell fulfills the organism's needs and use evidence to support that explanation, for example, to explain how and where DNA replication occurs and why this is needed for the organism's functions. Students could also be asked to coordinate information from various sources and argue for an interpretation, including the reasons that they do not accept a source that disagrees with their interpretation. Tasks teachers have typically assigned to students—either in class, for homework, or for assessment purposes—need to be carefully reconsidered in light of the learning goals of the *Framework* and the NGSS.

Expecting Instruction to Change Overnight

Shifting instruction to incorporate all of the scientific and engineering practices and designing tasks for students that integrate the three dimensions (practices, crosscutting concepts, and core ideas) will take time. Teachers first need to understand the changes expected and the reasons for them and then move in steps to incorporate these changes into their instruction. Not everything can be changed at once, nor will the first steps necessarily engender the sense of success that would foster commitment to the change. It is likely that 2-3 years of professional development for teachers will be needed to help them make the changes to instruction that are called for in the NGSS. Teachers will then need ongoing support to continue to refine their instructional practices. One approach for this kind of support might be participation in a teacher learning community devoted to this goal.

Expecting Teachers to Do It Alone

Teachers have considerable demands on their time and significant personal investment in the teaching strategies and materials that they have developed over time. Both of these factors often result in even the best teachers working in isolation. A more collaborative teaching culture is a necessary part of achieving the needed change.

The task of revamping an entire curriculum should not rest on the shoulders of a single teacher. At the same time, it is important to tap teachers' expertise and leadership abilities. To bring about the change that is embodied in the *Framework* and the NGSS, implementation needs to be structured to develop collaborative networks of teachers and school leaders within and across grades, buildings, districts, and states that work toward a shared vision (Cohen, 2011). Such collaboration can help teachers let go of lessons and units that may have been thoughtfully developed for previous standards and assessments, but that do not meet the expectations of the NGSS or are misaligned to the grade-level curriculum scope and sequence that is being implemented in their district.

Teachers can learn not just through their own experiences, but also through those of other teachers, including some who are in schools or districts that are further along the path to implementation of the NGSS. Teachers can support one another by sharing effective strategies or by collaborating to develop new units of instruction aligned to new scope and sequence expectations and to engagement of students in the practices. Students can benefit from experiencing a common culture of science learning across their school and across the school system. It is the job of leaders at all levels to create conditions that reduce isolation and facilitate cooperation and collaboration among science teachers.

Being Reluctant to Let Go of Familiar Units or Favorite Activities

Every teacher of science has a repertoire of ideas and activities that they have found effective in teaching. Each of these, while possibly still useful, will require reexamination, possible redesign, or even elimination in order to ensure that instruction is aligned to the performance expectations of the NGSS and engages students in science and engineering practices in ways that reflect the vision of the *Framework*. If lessons that were built for past standards do not support learning that combines all three dimensions of the *Framework,* (practices, crosscutting concepts, and disciplinary core ideas), and cannot be easily adapted, they may need to be dropped or replaced.

The *Framework* and the NGSS focus on developing fundamental science ideas at a deep conceptual level, which likely will involve pruning some of the details that teachers have frequently covered. Some science teachers have developed a wide variety of mnemonics and other creative solutions to support students in learning some of the specific facts that are not in the NGSS. It may be especially difficult for some teachers to leave out part of the curriculum that they have previously thought to be essential in favor of more time for deeper engagement in the core ideas and crosscutting concepts in the NGSS.

4

TEACHER AND LEADER LEARNING

RECOMMENDATION 5 Begin with leadership. State, district, and school leaders should designate teams that include teachers to lead implementation of the Next Generation Science Standards. Initial professional development efforts should be focused on these leadership teams. Team members should then be engaged in continuing professional learning appropriate to their roles to lead implementation of the necessary changes in curriculum, instruction and assessment.

RECOMMENDATION 6 Develop comprehensive, multiyear plans to support teachers' and administrators' learning. State, district, and school science education leaders should develop comprehensive multiyear plans for professional learning opportunities for teachers and administrators. These plans need to balance existing resources, meet expectations for milestones in implementation of the Next Generation Science Standards (NGSS), and take advantage of available tools and partners. The plans should take the needs of both current and new teachers into account and allow for ongoing refinement as schools and teachers gain expertise in implementing the NGSS.

RECOMMENDATION 7 Base design of professional learning on the best available evidence. When designing professional learning experiences, district and school leaders and providers of professional development should build on the key findings from research. Professional development should (1) be content specific; (2) connect to teacher's own instructional practice; (3) model the instructional approach being learned and ask teachers to analyze examples of it; (4) enable reflective collaboration; and (5) be a sustained element of a comprehensive and continuing support system. For sustained implementation, research shows that principals' understanding of and support for instructional change is key.

RECOMMENDATION 8 Leverage networks and partners. Science education leaders at the state and district level, and lead teachers should take full advantage of and cultivate partnerships with other districts, professional development networks, web-based professional development resources, science education researchers, and science-rich institutions—such as higher education institutions and science technology centers—to facilitate high-quality professional development.

LEADERSHIP

An important first step for implementing the Next Generation Science Standards (NGSS) is identifying science leadership teams that will provide a core element of the implementation and professional development. Cultivation of leaders for science starts by identifying teachers and administrators at the elementary, middle, and high school levels who have experience in science teaching and leadership capacity and who have a demonstrated interest in deepening their expertise in the new directions of the NGSS. Team members then need support for their own ongoing professional learning, as well as the responsibility to plan for and organize the professional learning of other teachers. They also need the authority, resources, time, and access for this work and to support and mentor other teachers. See Box 4-1 for a description of the importance of teacher leaders in improving science and mathematics instruction.

BOX 4-1

TEACHER LEADERS IN SYSTEMIC REFORM

In 1995, the National Science Foundation (NSF) initiated the Local Systemic Change Through Teacher Enhancement Program. The initiative's primary goal was to improve instruction in science, mathematics, and technology through teacher professional development within schools or school districts. By 2002, NSF had funded 88 projects that targeted science or mathematics (or both) at the elementary or secondary level (or both). The Local Systemic Change (LSC) projects were designed for all teachers in a jurisdiction; each teacher was required to participate in a minimum of 130 hours of professional development over the course of the project. The LSC Program also emphasized preparing teachers to implement district-designated mathematics and science instructional materials in their classes (Banilower et al., 2006).

In addition to providing professional development for teachers, the LSC Program promoted efforts to build a supportive environment for improving science, mathematics, and technology instruction. LSC projects were expected to align policy and practice within targeted districts and to engage in a range of activities to support reform. Those activities included

- building a comprehensive, shared vision of science, mathematics, and technology education;
- conducting a detailed self-study to assess the system's needs and strengths;
- promoting active partnerships and commitments among an array of stakeholders;
- designing a strategic plan that includes mechanisms for engaging teachers in high-quality professional development activities over the course of the project; and
- developing clearly defined, measurable outcomes for teaching and an evaluation plan that provides formative and summative feedback.

Evaluators of the LSC projects and the project directors concluded that teacher leaders were an essential component of success. Teachers on special assignment and school-based teacher leaders often assumed active roles on school and district committees during the initiative, and many continued in these roles after the NSF support ended. The project directors frequently attributed major project successes to teacher leaders, including efforts to align district curriculum with state and national standards, adopt high-quality instructional materials, and develop aligned assessments. Evaluators reported that teacher leaders' participation on reform-oriented committees helped broaden their understanding of district policies and practices and provided them with a new perspective on how change happens. According to one evaluation: "[The teacher leaders] are a dynamic group that is likely to influence policies and practices for years to come" (Banilower et al., 2006, p. 81).

At the start, there will be a need for focused professional development, likely from outside providers or networks, first for the administrators and teachers who will form the science leadership team, and then progressively involving all teachers who teach science at any level. Teachers need support to find and take advantage of the best available professional learning opportunities, both locally and online, to develop their instructional vision and practice. Administrators' understanding of and support for the changes in science instruction and learning goals is essential, particularly at the elementary level. Teachers need ongoing support beyond the first year of implementation to integrate the changes into their teaching style and instructional decision making (Allen et al., 2011; Martin and Hand, 2009; Ratcliffe et al., 2007). When significant new curriculum resources are added, professional development that is tightly linked to effective use of those resources will be important.

Professional development should incorporate discussion of the relationships between changes in science teaching expectations and changes in other subject areas, especially for the elementary grades, where most teachers teach multiple subjects. Understanding these relationships will allow teachers to take advantage of the synergies between science, mathematics, and English language arts by supporting development of students' skills across the curriculum in the context of science learning activities.

Even once the transition to the new standards is "complete," that is, everyone thinks they are doing what is needed, teachers and leaders should continue working to improve their understanding of the NGSS and of how best to support students' learning as described in the NGSS. One critical element to support this learning culture is ongoing opportunities for teachers to participate in learning communities facilitated by well-informed teacher leaders, with time to discuss and reflect on science instruction. It is also important to coordinate these kinds of conversations across grades and across disciplines (in the case of middle and high school teachers) on a regular basis. In situations where it is difficult to bring teachers from multiple grades or multiple science disciplines together, technology can be used to support ongoing collaboration.

LEARNING OPPORTUNITIES FOR TEACHERS[1]

Wilson (2013, p. 310) suggests that "the U.S. PD [professional development] system is a carnival of options" that is often not well matched to teachers' needs. Realigning professional development resources toward more effective and sustained approaches is essential for effective implementation of the NGSS.

Studies of professional development programs reveal an emerging consensus about the features that are most promising for supporting teacher learning. Those features, discussed in detail below, are a focus on specific content, connection to classroom practice, active learning, collaboration, and being sustained (Banilower et al., 2007; Borko, 2004; Garet et al., 2001; Heller et al., 2012; Penuel et al., 2009; Putnam and Borko, 2000; Roth et al., 2011; Wilson, 2013; Yoon et al., 2007). There have been only a few studies that specifically examined professional development in science (Heller et al., 2012; Penuel et al., 2009; Roth et al., 2011). Those studies support the findings of the more general studies and also show that professional development focused on students' thinking and analysis of instruction is more effective than professional development focused only on improving teachers' content knowledge in science (see Box 4-2).

A Focus on Specific Content

Professional development needs to be deeply connected to specific content (Garet et al., 2001). In the NGSS, content includes all three dimensions: practices, crosscutting concepts, and disciplinary core ideas. Professional learning opportunities should be designed such that teachers grapple with both the science itself and how students think and learn about that science. Interventions that focus primarily on deepening teachers' knowledge of disciplinary core ideas are likely to be insufficient. While knowledge of the science itself is essential and lack of such knowledge may pose challenges for teachers (Kanter and Konstantopoulos, 2010), such knowledge is not sufficient for teachers to be able to translate what they have learned into effective lessons for students (Heller et al., 2012). Teachers' knowledge of how to support student learning typically draws on general principles about learning (e.g., the importance of building on students' prior conceptions), but it critically depends on understanding those general principles in the context of specific disciplinary core ideas (e.g., the nature of matter) and recognizing the

[1]This section is based on a paper by Reiser (2013) written for the Invitational Research Symposium on Science Assessment convened by ETS in September 2013.

BOX 4-2

EXAMPLES OF SUCCESSFUL PROFESSIONAL DEVELOPMENT PROGRAMS IN SCIENCE

Simply telling teachers about the new standards or focusing solely on improving their science content knowledge is unlikely to lead to the kinds of sustained changes in instruction that will be needed to support the NGSS. Instead, science teachers need opportunities to examine students' thinking and analyze instruction. Two recent studies of professional development offer examples of these kinds of learning opportunities for science teachers.

In a large-scale study of 270 elementary teachers in 39 school districts across 6 states, Heller et al. (2012) compared four professional development courses for elementary teachers. All four courses involved the same science content; they differed in the ways they incorporated analyses of students' thinking and analyses of instruction. Each of the 4 intervention models involved 24 hours of contact time divided into eight 3-hour sessions:

- In one intervention model, teachers discussed narrative descriptions of extended examples from actual classrooms, which included student work, classroom discussions, and descriptions of the teachers' thinking and behavior.
- In a second intervention model, teachers examined and discussed their own students' work in the context of ongoing lessons.
- In the third intervention model, teachers engaged in reflection and analysis about their own learning as they participated in science investigations: they considered which ideas could be learned through the investigation, tricky or surprising concepts, and implications for students' learning.
- The fourth course served as a control group and involved only science content.

challenges that students frequently face in making sense of the particular new content ideas (Putnam and Borko, 2000).

Similarly, professional development to introduce science practices should not just provide generic guidance about how to support argumentation or how to help students develop science models. Instead, the practices are best developed, for both teachers and students, in the context of particular core ideas. For example, teachers need to be able to help students develop explanatory accounts of phenomena using the particle model of matter or evidence-based arguments about population biology or to design devices to minimize or maximize the transfer of thermal energy. The specific subject area lends context to the practice at the same time it

All four intervention models improved both teachers' and students' scores on tests of science content knowledge more than the scores of teachers and students in the control group. In addition, the effects of the intervention on teachers' students were stronger in the follow-up year than during the year of intervention.

The Science Teachers Learning from Lesson Analysis (STeLLA) project featured video-based analysis of instructional practice aimed at upper elementary teachers. The year-long professional development experience for teachers focused on how to create a coherent science storyline for students and how to elicit, support, and challenge students' thinking about specific science concepts. The study involved 48 teachers: 32 participated in the STeLLA program and 16 participated in a content-only program (Roth et al., 2011):

- Both groups participated in a 3-week summer institute focused on science content.
- The STeLLA participants also engaged in video analysis during the summer and in follow-up sessions during the year; they met in small groups facilitated by a program leader to discuss video cases. Teachers began with cases from unfamiliar teachers and later discussed videocases based on their own classrooms.
- The lessons of the STeLLA teachers were analyzed to determine whether they were using the strategies related to creating storylines and supporting students' thinking.

Both the STeLLA teachers and the teachers in the comparison group showed gains in science content knowledge, but the STeLLA teachers made greater gains. In addition, videotaped samples of lessons from the STeLLA teachers' classrooms show that by the end of the year they were implementing many of the strategies related to supporting a science content storyline and supporting students' thinking. Students of teachers who participated in the program showed greater learning gains in the year after the teachers' participation than students in the year previous to the teacher's participation.

enriches teachers' understanding of how student engagement in these practices facilitates and deepens student learning.

Connected to Teachers' Instructional Practice

Learning opportunities for teachers need to be connected to issues of teachers' own practice (Ball and Cohen, 1996; Borko, 2004; Garet et al., 2001; Roth et al, 2011). Teachers need opportunities and support to begin to apply the ideas in their own practice (Darling-Hammond, 1995; Putnam and Borko, 2000) and then to discuss with mentors or colleagues the challenges that arise. In the vast majority of cases, those discussions have to include an explicit focus on both specific content (which includes all three dimensions of the *Framework*) and specific instructional materials.

Because the issues and professional development needs of elementary, middle, and high school science teachers are different, districts will have to plan separately for each level, while recognizing when and how to facilitate conversations and reflection on progress across these levels. Even within the elementary level, teachers at different grades are responsible for different science standards and may benefit the most from professional development that is grounded in topics that they are expected to teach. It is important to help teachers at all grade levels identify ways to support students' reading and writing about science and their use of mathematics and computational thinking in science. However, the needs of elementary teachers who typically teach mathematics and English language arts as well as science may be different from those of secondary teachers who do not.

As the implementation of the NGSS progresses, teachers' need for professional development opportunities does not end, but the type of opportunity that will be most useful to them changes. Ideally, every teacher would have an individual professional learning plan; short of that, there needs to be a rich but coherent menu of professional development opportunities. Some aspects of the professional development menu should consider teaming teachers across schools and districts for focus on a particular grade level. Another possibility is teaming teachers across two or more grades that address similar core ideas, at different levels, so that teachers can connect their own part with what comes before or after in the students' learning trajectory for that topic.

Active Engagement

Professional development tasks need to involve teachers in active reflection and problem solving (Garet et al., 2001). Teachers, like all learners, need to go beyond being presented with ideas and strategies: they need opportunities to analyze specific problems or situations and to figure out what strategies to apply. In professional development, this approach translates into opportunities to study examples of classroom interaction that reflect a particular teaching and learning issue, such as eliciting students' models and model-based explanations, helping students develop and defend arguments based on evidence, facilitating engineering design, or selecting tasks that can also be used to formatively assess students' thinking. Such examples can be used as material for analysis and discussion, rather than "model examples" of routines to be followed. In one study, teachers who participated in professional development that included intensive analysis of classroom-based video cases of particular teaching moments along with a focus on the subject matter learned more and produced more learning gains for their students than teachers involved only in professional development on the science ideas alone (Roth et al., 2011).

Teachers need to analyze and deconstruct teaching examples in order to figure out what can be applied to their own classrooms. They need to work with rich cases that reflect the complexity of the learning desired and contain enough context to explore the rationale for student-student and student-teacher interactions that occur and to track their changes over time (Borko, 2004). Rich cases also provide examples in which teachers can explore what types of tasks can provide experience with phenomena, raise questions, and help students construct explanations to make sense of the target ideas (Ball and Cohen, 1996, 1999; Borko et al., 2008; Roth et al., 2011).

Collaboration

Learning experiences for teachers should be collaborative and support teachers in working together to understand, apply, and reflect on implementation of the NGSS (Garet et al., 2001; Wilson, 2013). Such collaboration is a key strategy for teachers to continue to deepen and refine their understanding of the NGSS (Putnam and Borko, 2000). Collaborative analysis and discussion of specific examples of practice can create opportunities for the analysis needed to dig beneath the surface characteristics of the NGSS and to explore substantive issues in applying the standards in practice (Sherin and Han, 2004; van Es and Sherin, 2007). In investigating cases of science teaching, teachers could work together to debate their interpretations and consensus as they do the science activities themselves, analyze student work, and analyze teaching interactions. This kind of collaboration also develops teachers' understanding of the importance of collaboration for their students. Teachers also need supportive colleagues and particularly school administrators who understand the needed changes to persist in implementing new strategies learned in professional development programs.

Sufficient Time

Successful professional development programs require sufficient investment of time to enable teachers to grapple with new ideas, analyze examples of the ideas in action (such as student work or records of classroom interactions), and make step-by-step progress in understanding and applying the new ideas. Repeated experiences are needed to enable teachers to successfully integrate new elements into their teaching practice and use them flexibly.

Effective professional development programs involve extended sessions, including some that are spread across time, such as long intensive workshops during the summer with follow-up sessions during the school year. A typical program might consist of eight to ten 3-hour sessions in the summer (see Heller et al., 2012) or even more intensive interventions, such as more than 60 hours in the summer followed by 30 hours spread over the school year (Roth et al., 2011). One-shot professional development programs are unlikely to be effective for helping teachers change their instruction (Yoon et al., 2007).

A Coherent and Ongoing System of Support

An issue that emerges as critical to changes in practice is the need for alignment of professional development with other components of the system such as curriculum or assessment (Garet et al., 2001). Different aspects of coherence have been highlighted across studies of professional development—coherence with the teachers' and principal's goals, alignment with changes in standards, alignment with assessments, and curriculum materials that reflect the reforms (Darling-Hammond, 1995; Wilson, 2013).

To support teacher learning as part of implementing the NGSS, then, connecting to teaching practice requires that teachers explore what a coherent system of student learning, classroom instruction, assessment, and curriculum materials needs to achieve, and work on coordinated changes across these corresponding parts of a system. Teachers (and their supervisors) need to recognize that they and their students will continue to change over multiple years of implementation. For high school teachers, it may be as many as 10 years before the majority of entering students arrive well prepared for the new curriculum that the teachers are expected to implement. Thus, teachers will need to continue to refine their approach and their expectations of what students can do over many years.

LEARNING OPPORTUNITIES FOR ADMINISTRATORS

In order for school leaders and district administrators to understand the needs of science teachers, administrators will need sufficient professional development to recognize what is and what is not aligned to the new vision and productive for the NGSS learning outcomes. Administrators will need an opportunity to experience the type of science learning envisioned by *A Framework for K-12 Science Education: Practices, Crosscutting Concepts, and Core Ideas* (National Research

BOX 4-3

THE IMPORTANCE OF ADMINISTRATORS

Directors of the Local Systemic Change (LSC) projects (see Box 4-1) typically cited support by principals as the most important factor in determining teacher participation in professional development and in developing a supportive context for reform at the school level. Evaluators and project directors cited examples of principals who were active participants in professional development and who looked for ways to support teacher leaders, budget new resources, create opportunities for teacher collaboration, and educate parents about new mathematics and science programs. As summarized in the Capstone report on the initiative (Banilower et al., 2006, p. 88):

> [I]n many ways, principals played key roles in determining the outcomes of [projects]—from encouraging teachers to participate in professional development, to supporting teachers' use of high-quality materials and inquiry-based practices, to enabling the work of teacher leaders, to making time for teachers' to participate in site-based professional development.

Where LSC projects established strong working relationships with district administrators—including superintendents, school boards, curriculum directors, and others—project directors and evaluators noted that the potential for sustained support increased significantly. Superintendents were integral to removing roadblocks and demonstrated their commitment by using general funds to adopt new materials after they were taken off the state adoption list. Other superintendents who attended national leadership institutes with leaders of the initiative often demonstrated high levels of commitment by promoting the adoption of designated instructional materials or mandating participation by teachers or principals in professional development (Banilower et al., 2006).

Council, 2012) and the *Next Generation Science Standards: For States, By States* (NGSS Lead States, 2013) and to discuss with others what this means for their schools and the teachers teaching science within them. See Box 4-3 for a discussion of the importance of administrators.

PITFALLS TO AVOID

Underestimating the Shift Needed in One's Own Practice

Teachers, administrators, and professional development providers who view the *Framework* and the NGSS through the lens of current practice may underestimate the need for change (Spillane et al., 2002). For example, physics teachers may consider their students are already learning engineering by building and testing model bridges or conducting egg-drop contests. However, these activities do not necessarily represent the NGSS-aligned engineering instruction unless they have been carefully designed to incorporate engineering practices (such as defining problems in terms of criteria and constraints) and involve students in building, extending, or using scientific concepts as part of the engineering project (such as forces or transfer of energy). Similarly, many teachers or administrators may see the science and engineering practices as essentially equivalent to "inquiry" with just a new name or equivalent to teaching "the scientific method" (Reiser, 2013; Windschitl et al., 2008). Such views miss the NGSS's emphases on knowledge building, social interaction and discourse, analysis, and reasoning as part of scientific and engineering practices.

It will be easy to underestimate the degree of change in instructional practice needed in order to engage students in the practices of science and engineering. Existing activities for students may have the appearance of engagement in a science or engineering practice because they are hands on or involve students designing experiments, but they may miss the critical aspects of building and testing explanatory ideas.

Some elements of the *Framework* and the NGSS are already being implemented in some classrooms. Indeed, the vision of the *Framework* and the NGSS is built on a firm foundation of classroom-based research about what is most effective for science learning. However, few classrooms have been implementing the full range of practices. In addition, the sequencing of core ideas across grades requires some rethinking of what is taught when. Simply doing a check-the-box alignment of old standards, curricula, curriculum materials, or assessment tasks that matches them to pieces of the new standards will not be sufficient for implementing the vision embodied in the NGSS.

Underestimating the Need for Ongoing Support

Teachers need time to practice, and they need ongoing reinforcement to support the effort it takes to change both their own teaching practice and their classroom culture. It takes sustained effort and ongoing learning and reflection for any teacher to achieve facility and flexibility in implementing a new approach to instruction. Support can take the form of mentoring, more and different professional development opportunities, or time for participating in a professional learning community. Support for teachers also requires that school and district leaders have themselves received appropriate professional development about the NGSS and thus share the vision of teaching and learning and have the knowledge to appropriately respond to teachers' practice and learning needs.

Failing to Provide Opportunities for Administrators to Learn About the NGSS

School administrators who do not understand the nature of the changes required by the NGSS may place demands on teachers, including criteria for evaluating teachers, which undermine implementation of the new strategies needed. It is unrealistic, for example, to assume that each day or two the classroom should move onto a different performance expectation or to assume one-to-one mappings between sequences of lessons and performance expectations. Multiple lessons will need to build toward performance expectations over time (Krajcik et al., 2014). It is essential that administrators themselves learn about the goals and strategies to meet them that are implied by the adoption of the NGSS so that they recognize the changes that teachers are attempting and support teachers in implementing them effectively. Opportunities for administrators to become familiar with the *Framework* and the NGSS will need to be provided by districts and states.

Offering "One Size Fits All" Learning Opportunities

Different teachers at different grade levels have different needs. Even at a given level, some teachers have stronger science backgrounds than others or are at different places along the path toward teaching aligned to the vision of the *Framework* and the NGSS. A common complaint of teachers is that their districts require them to "waste their time" attending professional development that is directed to skills they already have or generic teaching strategies not well matched to the subject matter that they teach. It is critical that district leaders ensure that their professional development opportunities are structured to make effective use of teacher time and meet the teachers' needs. In general, this approach will require offering a menu of options and giving teachers some choices about how best to meet their professional development needs.

5

CURRICULUM MATERIALS

RECOMMENDATION 9 Do not rush to completely replace all curriculum materials. States, districts, and schools should not rush to purchase an entirely new set of curriculum materials since many existing materials are not aligned with the Next Generation Science Standards (NGSS). Until new materials are available, district leadership teams in science will need to work with teachers to revise existing units and identify supplemental resources to support the new vision of instruction. In searching for supplemental materials, district leaders and teachers should look for those designed around goals for student learning that are consistent with the NGSS.

RECOMMENDATION 10 Decide on course scope and sequencing. State and district leaders will need to make decisions regarding the scope and sequence of courses in science. Scope and sequence is especially important for grades 6-12, for which the performance expectations of the Next Generation Science Standards (NGSS) are organized in grade bands (6-8 and 9-12). The process of planning scope and sequence should be guided by the strategies outlined in Appendix K of the NGSS.

RECOMMENDATION 11 Be critical consumers of new curriculum materials. District leaders should plan to adopt and invest in curriculum materials developed for the Next Generation Science Standards (NGSS) when high-quality materials become available and in keeping with their own curriculum adoption schedule. District leadership teams should use a clear set of measures and tools with which to judge whether curriculum materials are truly consistent with the goals of *A Framework for K-12 Science Education: Practices, Crosscutting Concepts, and Core Ideas* and the NGSS. Individuals involved in the adoption process should be trained to use those measures and tools.

RECOMMENDATION 12 Attend to coherence in the curriculum. Curriculum designers and curriculum selection teams should ensure that curriculum materials are designed with a coherent trajectory for students' learning. The performance expectations in the Next Generation Science Standards (NGSS) are the target outcomes for the end of a grade level or grade band, and curricula will need to elaborate on a sequence of experiences that will help students meet those expectations. Students need to experience the practices in varied combinations and in multiple contexts to be able to use them as required to meet the NGSS performance expectations.

THE ROLE OF CURRICULA

A set of standards is not a curriculum; rather, it defines the outcomes expected for students from the enacted curriculum. In the Next Generation Science Standards (NGSS) these outcomes are framed as performance expectations that include practices as well as disciplinary core ideas and crosscutting concepts. However, because the practices work together in coherent investigations or engineering projects, working toward the performance expectations typically requires engaging students in more combinations of disciplinary core ideas and practices than the combinations specified in the performance expectations (Krajcik et al., 2014). Thus, teachers need resources that articulate coherent trajectories of questions to investigate or problems to solve that bring together target core ideas, crosscutting concepts, and practices. Such resources may include text materials, online resources (such as simulations or access to data), materials and equipment for investigative and design activities, and teacher manuals that include supports for pedagogical strategies needed to enact these lessons. A fully developed curriculum may provide all of these as a single package, but often teachers will draw from multiple resources in designing their instruction.

Among the critical curriculum resources for science are time, space, equipment, and expendable materials that can be used for investigative and design projects (National Research Council, 2006b). Districts need to consider how their schedule, space use, and materials budgets can be designed to support the NGSS student goals or can impede achieving them. The need for these resources often makes providing quality science teaching more expensive than some other subjects. Sharing of equipment, materials supplies, and even space by several teachers can provide some economies in purchasing, but it requires a well-developed management and replenishment system to function well. One example of such sharing, at the elementary level, is kits on carts.

The NGSS describes the year-by-year sequence of standards for kindergarten through grade 5 and groupings of standards for grades 6-8 and 9-12. Explicit curriculum scope and sequence plans will need to go further, deciding how to sequence topics within each year or grade block and how to ensure that students engage in all the science and engineering practices and apply all the crosscutting concepts in multiple disciplinary contexts. For suggestions of grouping standards for middle and high school courses, see Appendix K of the NGSS.

Simply defining what standards are to be "covered" in a given year will not be sufficient. Coherence within a unit, between units across a year, and from one year to the next is key in engaging students in the type of knowledge building targeted in the NGSS (Fortus and Krajcik, 2012; Reiser, 2013). Curriculum units need to be crafted such that they present coherent investigations or engineering problems, in which questions and phenomena motivate building and using disciplinary and crosscutting ideas, This approach can be contrasted with simply sequencing topics in traditional sequences that make sense to experts but are unmotivated for learners (Krajcik et al., 2008, 2014). Curriculum units need to be sequenced across a year so that students can build ideas across time in coherent learning progressions, in which questions or challenges, gaps in models, and new phenomena motivate developing deeper disciplinary core and crosscutting ideas.

States and districts, in conjunction with science coordinators and educators, will need to decide on and implement course options to provide a coherent sequence of science instruction across the grades. They will also need to plan how to transition to a new sequence without creating large gaps in students' learning. This planning may involve consideration of how science course sequences coordinate with course sequences in career and technical education. Educators will then need to seek curriculum materials that match the chosen sequence of instruction. Teachers will need to understand their part in the multiyear scope and sequence and support students in building on their prior knowledge, while they learn new topics or deepen their understanding of those they have taught before. To do this, teachers need opportunities to communicate and collaborate both within and across grade levels and school levels.

As this report was being written, the committee was not aware of any yearlong, comprehensive curriculum resources at any grade level built explicitly for the NGSS, though a number were under development. Developing and phasing in a full set of new curriculum materials aligned with the NGSS will take time. There are, however, some existing research-based curriculum materials with evidence of impact on student learning that support students in science and engineering practices and address some of the disciplinary core ideas and crosscutting concepts from the *Framework* (see Sneider, 2015). Schools and teachers can work with these materials or adapt their existing materials. Efforts to use such research-based materials or selectively adapt existing units could help districts shift classroom teaching toward the vision of the *Framework* and the NGSS and help teachers develop a deep understanding of the NGSS. It will also help teachers and district leaders to be better able to evaluate the quality of more complete sequences of curriculum materials as they become available. Box 5-1 discusses the relationship of curriculum and professional development.

District leaders should coordinate collaboration among K-12 teachers to evaluate existing materials and lessons for how well they reflect all three dimensions of the NGSS. Instructional units on topics that are included in the standards for a grade level should be adapted to focus around student learning experiences that engage students in the science and engineering practices or replaced with research-based materials that do so, when available. These revisions may involve eliminating topics or units that are not included in the NGSS. They also may involve seeking units originally designed for other grade levels by refining and revising topics that were previously not included at that grade level.

Individual teachers should not be expected to redesign curriculum unaided. Participation in a group activity to redesign a particular unit can be an effective professional development opportunity (Penuel et al., 2011). When possible, the redesign team should include outside experts, including content-area experts,

BOX 5-1

PAIRING CURRICULUM MATERIALS WITH PROFESSIONAL DEVELOPMENT

Evaluations of the Local System Change initiative (see Box 4-1 in Chapter 4) show that lessons based on instructional materials designated by the district were more likely to be rated as high quality than lessons that were heavily modified by individual teachers (Bowes and Banilower, 2004). Use of district-designated instructional materials was positively correlated with several key outcomes, including more frequent use of investigative classroom practices and greater emphasis on important and developmentally appropriate mathematics and science.

Professional development also played an important role in supporting teachers to change their instruction: more lessons of teachers who participated in more hours of professional development were rated as high quality. In self-contained classrooms at the elementary level, participation in professional development was positively related to the number of hours spent on science instruction (Heck et al., 2006b).

The combination of professional development and the use of designated instructional materials appears to have had a greater effect than either factor alone (Bowes and Banilower, 2004). Also, as teachers participated in more professional development, their use of the district-designated materials increased (Heck et al., 2006b).

experts in curriculum development, and experts on the effective implementation of science practices in the classroom. That expertise can considerably enrich the outcome of a redesign effort and the teacher learning that occurs through the work.

DEVELOPING NEW MATERIALS

Designing a quality set of curriculum resources for a new course or course sequence is a demanding multiyear, multi-expert team process. In designing, development teams need to include experts in science, science learning, assessment design, equity and diversity, and science teaching, each at the appropriate grade level (National Research Council, 2014a). Those working to develop new resource materials, instructional units, and comprehensive curricula based on the *Framework* and the NGSS will need to ensure that student tasks and assessment activities in the materials (such as mid- and end-of-chapter activities, suggested tasks for unit assessment, and online activities) mirror the integration of the dimensions that are expected in the vision of instruction.

The curriculum materials will need to include support for teachers to use the formative assessment process to gather information about student learning in all three dimensions (National Research Council, 2014a). Developers need to recognize that most traditional approaches to curriculum materials, in which teachers or expository text present new ideas first, and then students apply them in labs or exercises, do not reflect the three dimensions of the NGSS, in which students engage in the science and engineering practices to develop and use the disciplinary core and crosscutting ideas with guidance from teachers. Attention to the practices students are expected to engage in and the ways in which teachers can support students in developing these capabilities are as important as attention to the disciplinary topics and ideas that are to be learned. Curricula will need to achieve a new balance between time spent in the productive struggle to investigate and explain phenomena or to design problem solutions and the number of topics addressed (see, e.g., Southwick, 2013).

The curriculum designers also need to consider multiple dimensions of diversity and how to connect with students' cultural and linguistic resources. Although designers are used to taking account of these issues in text and perhaps also in the pictures included with the text, the issue of building instruction around real-world phenomena and design tasks around real-world problems adds a new dimension to this issue, which provides both new opportunities and new challenges for sensitivity to equity and access concerns (Lee et al., 2014). For guidance on supporting diverse populations, see Appendix D of the NGSS.

SCHOOLS, DISTRICTS, AND TEACHERS AS CRITICAL CONSUMERS

Eventually it will be time for a state to adopt or a district to purchase new science curriculum materials. Before doing so, they will need to have made course scope and sequence decisions about their middle and high school programs so that they can seek materials matched to their courses. To design a scope and sequence that also reflects the goals of the *Framework* and the NGSS, districts can find guidance in Appendix K of the NGSS.

Before actually selecting the materials to be purchased, school and district leaders should become critical consumers of curriculum materials on the basis of experience of what it means to teach science that meets the vision of the NGSS learning goals. They should approach the adoption of texts and planning of units or lessons with a clear set of criteria for consistency with the vision of the *Framework* and the NGSS. Evaluation processes for curriculum materials that only look at what content topics are included in the materials will not be useful. Rather, the "content" needs to include all three dimensions of the NGSS, and they must be developed in a coherent fashion so that the resources support instruction that meets the vision of the *Framework* and the NGSS. For example, textbooks that include all possible topics rather than focusing on the disciplinary core ideas should not be selected for use. Similarly, textbooks should not be selected that include the disciplinary core ideas but do not include approaches that have students engaged meaningfully in the science and engineering practices to develop and use those disciplinary core ideas.

Some school districts are moving toward use of open access materials rather than undertaking traditional textbook adoption. But use of open access materials also needs to be guided by the vision for science learning in the *Framework* and the NGSS and a clear set of criteria for consistency with the vision. The materials will need to be carefully sequenced to support students' developing understanding of the core ideas and crosscutting concepts.

A claim that a curriculum is "aligned" to the NGSS does not mean that it was designed specifically for the NGSS and fully incorporates the practices and crosscutting concepts and follows the progressions of core ideas. It is likely, as has occurred with Common Core State Standards, that many of the most rapidly available textbooks and related resources claiming alignment to the NGSS will be superficially rather than deeply aligned and will not have been substantially redesigned (see Herold and Molnar, 2014).

Without clear criteria for making choices, districts could spend significant money on materials that are not what they really need. The selection process can be facilitated through the use of tools that support a systematic evaluation that goes beyond judging superficial alignment and take account of inclusion of all three dimensions (practices, crosscutting concepts, and core ideas) with coherent sequencing. For example, working with science educators, administrators, and experts in science learning, Achieve developed the EQuIP Next Generation Science Standards Rubric for use as a tool in selecting curriculum materials. (The EQuIP NGSS Rubric is available online.) Such a tool can form a useful starting point for states and districts as they develop their own evaluation tools.

Teacher leaders can be valuable participants in the process of identifying quality curriculum materials that are consistent with the *Framework* and the NGSS. In order to be able to evaluate curriculum resources, it will be essential for teacher leaders to have experience with modifying existing lessons or units for the NGSS or designing and implementing new ones.

Teachers often use curriculum resources from multiple sources to design and support their own units for teaching particular topics. As teachers redesign their units in the context of the NGSS, they should consider whether there are new resources available that better support instruction that follows the vision of the *Framework* and the NGSS. To do so, they need good evaluation tools and procedures, just as in the case of larger scale district adoption and purchase of materials.

PITFALLS TO AVOID

Asking "Which Standard Are You Teaching Today"?

A "standard by standard" approach to curriculum does not work for the NGSS. The NGSS are student performance outcomes for the end of a grade level or grade band; they are not a list of activities for the classroom. Covering standards one at a time would lead to redundancies and fragmented learning. The particular combinations of the three dimensions represented in the NGSS are not prescriptive of how they should be combined in instruction. Facility with any one practice requires using others, and all of them need to be experienced in the context of learning multiple different core ideas. That is the way that students can gain facility in using them in any particular context in a testing situation. Moreover, in order to provide time for students to undertake investigations and engage in discourse, there is insufficient time to address each standard separately. Instead, standards will need to be bundled into instructional units that recognize the interconnections between the science and engineering practices, subideas within and across disciplinary core ideas, and the role of the crosscutting concepts in elucidating these connections (Krajcik et al., 2014; Pruitt, 2014).

Waiting Before Beginning to Change Instruction

To be able to evaluate whether or not curricula actually meet the expectations of the NGSS, it will be important for educators to experiment with trying some of the instructional shifts *before* selecting or developing curricula. Having teacher teams reevaluate existing materials, explore potential materials, and work strategically to adapt particular units of instruction to align with the NGSS will help build capacity for teaching in ways that align with the NGSS. Without this depth of experience, teachers will not be prepared to recognize curricula that do a good job of incorporating the three dimensions of the *Framework* throughout student learning.

Failing to Provide Resources to Support Students' Investigations and Design Projects

Curriculum materials are not the only resources that teachers need in order to implement the NGSS vision of instruction. Every investigation or engineering design project requires space, equipment, and resources, whether it is a laboratory-type investigation, a field study conducted in the school yard, or an engineering project conducted in the classroom. Other significant resources need to be considered, which include storage and preparation space, supplies, equipment for measurement and data collection; appropriate access to computers and software; availability of classroom space; and a master schedule that supports work on projects over time (National Research Council, 2006b).

6

ASSESSMENT

RECOMMENDATION 13 Create a new system of science assessment and monitoring. State science education leaders should create a long-term plan to develop and implement a new system of state science assessments that are designed to measure the performance expectations in the Next Generation Science Standards. The system should incorporate multiple elements, including on-demand tests, classroom-embedded assessments, and measures of opportunity to learn at the state or district level. When possible, state science education leaders and those responsible for state assessment should consider developing partnerships, perhaps with other states, to facilitate the work of developing new science assessments.

RECOMMENDATION 14 Help teachers develop appropriate formative assessment strategies. School leaders need to ensure that professional development for science teachers covers issues of assessment and supports teachers in using formative assessment of student thinking to inform ongoing instruction.

A SYSTEM OF ASSESSMENT

The Next Generation Science Standards (NGSS) describe specific goals for science learning in the form of *performance expectations,* statements about what students should know and be able to do at each grade level. Each performance expectation incorporates a practice in the context of a core idea and may also require students to call on a particular crosscutting concept.[1]

The performance expectations place significant demands on science learning at every grade level. It will not be feasible to assess each performance expectation for a given grade level during a single assessment occasion. Students will need multiple assessment opportunities—using a variety of formats—to demonstrate that their competence meets the expectations for a given grade level.

Measuring the performance expectations in the NGSS and providing all stakeholders—students, teachers, administrators, policy makers, and the public—with the information each needs about student learning will require assessments that are different in key ways from current science assessments. Specifically, the tasks designed to assess the performance expectations in the NGSS will need the following characteristics:

- include multiple components that reflect the connected use of different scientific practices in the context of interconnected disciplinary ideas and crosscutting concepts;
- address the progressive nature of learning by providing information about where students fall on a continuum between expected beginning and ending points in a given unit or grade; and
- include an interpretive system for evaluating a range of student products that are specific enough to be useful for helping teachers understand the range of student responses and, for formative assessment, provide tools that can help teachers decide on next steps in instruction.

The National Research Council (2014a) report on assessment provides numerous examples of the kinds of assessment tasks that have these characteristics.

Building on the advice in previous National Research Council reports (2006a, 2012, 2014a), the committee recommends a systems approach to science assessment. The system should include a range of assessment strategies that

[1]For a detailed discussion and analysis of assessment, see National Research Council (2014a), on which this chapter is based.

are designed to answer different kinds of questions for different stakeholders (students, teachers, administrators, policy makers, and the public). Such a system needs to include three components:

1. assessments designed to support classroom instruction,
2. assessments designed to monitor science learning on a broader scale, and
3. a series of indicators to track whether students are provided with adequate opportunity to learn science in the ways laid out in *A Framework for K-12 Science Education: Practices, Crosscutting Concepts, and Core Ideas* (hereafter referred to as "the *Framework*") and the NGSS.

The rest of this section discusses each of these components.

Assessment to Support Classroom Instruction

Classroom assessments are an integral part of instruction and learning and should include both formative and summative tasks. Formative tasks are those that are specifically designed to be used to guide instructional decision making and lesson planning; summative tasks are those that are specifically designed to assign student grades.

Assessments Designed to Monitor Science Learning on a Broader Scale

Student learning needs to be monitored over time in order to evaluate the effectiveness of the science education system. Given the breadth and depth of material covered in the NGSS, new approaches will be needed to monitor students' learning.

Assessments will need to include a variety of response formats, one of which needs to be performance-based questions that require students to construct or supply an answer, produce a product, or perform an activity. Although performance-based questions are especially suitable for assessing some aspects of student proficiency on the NGSS performance expectations, it will not be feasible to address the full breadth and depth of the NGSS performance expectations for a given grade level with a single external assessment comprised solely or mostly of performance-based questions. This is because performance-based questions take a significant amount of time to complete, and many of them would be needed in order to fully cover the set of performance expectations for a grade level. It will therefore be impossible to assess every student, on every standard, every year, using a one-time, on-demand test.

Thus, information from external on-demand assessments (those administered at a time mandated by the state) should be supplemented with information gathered from classroom-embedded assessments, which are administered at a time determined by the district or school or by the teacher, a time that fits the instructional sequence in the classroom. Classroom-embedded assessments may take various forms. They could be self-contained curricular units, which include instructional materials and assessments provided by the state or district to be administered in classrooms. Alternatively, a state or district might develop item banks of tasks that could be used at the appropriate time in classrooms. Another approach would be for states or districts to require that students in certain grade levels assemble portfolios of work products that demonstrate their levels of proficiency.

Indicators to Track Students' Opportunity to Learn

It is important to ensure that the dramatic changes in curriculum, instruction, and assessments prompted by the *Framework* and the NGSS do not exacerbate current inequities in science education. Instead, it is expected that the changes can begin to reduce inequities, while raising the level of science education for all students. Information should be routinely collected to monitor the quality of the classroom instruction that students receive, to determine whether all students have the opportunity to learn science in the way called for in the *Framework*, and to see whether schools have the resources they need to support science learning. This information might include onsite program inspections, student and teacher surveys, monitoring of teachers' professional development, and a system for periodic documentation of samples of teachers' lesson plans and associated student work (National Research Council, 2014a). Some observation of classroom instruction is important in order to ensure that the science and engineering practices are being implemented.

IMPLEMENTING A NEW ASSESSMENT SYSTEM

The systems approach to science assessment that we recommend cannot be reached by simply tinkering with an old system. A systematic but gradual process that reflects carefully considered priorities and timelines will be needed to make the transition to an assessment system that supports the vision of the *Framework*. Those priorities should begin with what is both necessary and possible in the short term while also establishing long-term goals for implementation of a fully integrated and coherent system of curriculum, instruction, and assessment. State leaders and educators should expect the development and implementation of the new system to take place in stages, over a number of years. Teachers will want to know the plans and timelines for changes in assessment at the state level; and at the same time, they will need professional development that supports them in using more open-ended assessment tasks in the classroom context.

The new system should be developed with an approach that begins with the process of designing assessments for the classroom, perhaps integrated into instructional units or curriculum materials, and then moves to designing large-scale assessments. Placing the initial focus on assessments that are close to the point of instruction will be the best way to identify successful ways to teach and assess knowledge of science practices as well as crosscutting concepts in specific disciplinary contexts. Effective strategies can then serve as the basis for developing assessments at other levels, including those used for accountability.

In designing and implementing assessment systems, states will need to focus on professional development. States will need to include adequate time and resources for professional development related to assessment strategies so that teachers can be properly prepared and guided and so that curriculum and assessment developers can adapt their work to the vision of the *Framework*.

State leaders who commission assessment development should ensure that the contracts address the changes called for by the *Framework* and the NGSS. They should therefore include in the contracts substantial amounts of time for the initial work and revision that will be needed to develop and implement such assessments: multiple cycles of design-based research will be necessary. Existing item banks are likely to be inadequate for gauging students' learning in alignment with the NGSS.

Existing and emerging technologies will be critical tools for creating a science assessment system that meets the goals of the NGSS, particularly those that permit the assessment of performance expectations that combine practices, core ideas, and crosscutting concepts. Technology will also be important for streamlining assessment administration and scoring.

States are likely to be able to capitalize on efforts already under way to implement the new Common Core State Standards in English language arts and mathematics, which have required educators to integrate technology for assessment along with new learning expectations and instruction. Nevertheless, the approach to science assessment recommended in the NRC report (National Research Council, 2014a) *Developing Assessments for the Next Generation Science Standards*, and that we endorse here, may require additional modifications

BOX 6-1

NEW ENGLAND COMMON ASSESSMENT PROGRAM

The New England Common Assessment Program (NECAP) is a series of reading, writing, mathematics, and science achievement tests, administered annually, that were developed in response to the federal No Child Left Behind Act. Students in New Hampshire, Rhode Island, and Vermont have been participating in NECAP since 2005, and Maine joined the program in 2009.

The state departments of education in New Hampshire, Rhode Island, and Vermont developed a common set of grade-level expectations and test specifications in mathematics, reading, and writing. The success of that effort led to development of common assessment targets and test specifications for science. Student scores are reported at four levels of academic achievement; Proficient with Distinction, Proficient, Partially Proficient, and Substantially Below Proficient. Reading and math are assessed in grades 3-8 and 11, writing is assessed in grades 5, 8, and 11, and science is assessed in grades 4, 8 and 11. The reading, mathematics, and writing tests are administered each year in October. The science tests are administered in May.

to current systems. States will need to carefully set their priorities and adopt a thoughtful, reflective, and gradual process for making the transition to an assessment system and technology platform for assessment that can support the vision of the *Framework*. Effective use of technology for both instructional and assessment purposes will be critical. For example, if the system includes classroom-based performance tasks, technology will be needed to allow teachers to submit student work products, share assessment rubrics, and grade the work of other teachers' students.

Given the complexity of the science assessment system envisioned, state science education leaders and those responsible for state assessment should consider partnerships with other states for the work. Multiple small coalitions may lead to a richer set of possibilities being developed than would be developed by only one or two large coalitions. An example of a relatively small coalition is the New England Common Assessment Program: see Box 6-1. The different coalitions could then share information with one another and with state leaders about their systems, blueprints for assessments, assessment tasks, and resources for assessment development. The goal should be to provide teachers and students with the best tools possible for assessing student learning in the classroom, as well to provide required accountability information through externally mandated assessments.

PITFALLS TO AVOID

Failing to Differentiate the Purposes of Assessment

Teachers collect assessment data for a variety of purposes. Classroom assessments are used to diagnose student needs near the beginning of a unit, to monitor progress along the way, and to find out how students are thinking about a topic so that teachers can determine how best to support students' learning or so students can evaluate their own progress. Assessments are also used to assign grades or to determine the effectiveness of a given unit for the class as a whole. It is important for a teacher to be clear from the start how the data will be used so that the right data can be collected and analyzed in a timely fashion. For example, for formative (classroom) assessments, ones that simply mimic external assessment tasks are unlikely to be useful (Penuel and Shepard, in press).

Failing to Respond to Assessment Results

There is no point in collecting assessment data if they are not used. In fact, collecting unnecessary data can be detrimental since assessment takes time away from learning experiences. Assessment data that are reported to teachers, or to students, too long after the assessment lose their effectiveness for supporting further student learning. It is also important for schools and districts to address any inequities that are revealed through the assessments of opportunities to learn.

Using Old Assessments While Mandating New Instructional Methods

It is unrealistic to expect teachers to immediately incorporate all the changes in instruction that are needed to support the NGSS. Instead, a staged approach, with ongoing professional development support, will be needed. Thus, it will be ineffective to continue to use old-style assessments to measure students and teachers while asking teachers to shift their teaching practice. Wherever possible, the transition needs to be supported by temporary relief from high-stakes accountability targets to allow both teachers and students the time to "hit their stride" with new demands of the NGSS (National Research Council, 2014a). The challenge for leaders is to find effective ways to monitor and support this progress while alleviating the anxiety of penalties for inadequate performance on tests not aligned to the new standards that can stifle attempts to make changes.

7

COLLABORATION, NETWORKS, AND PARTNERSHIPS

RECOMMENDATION 15 Create opportunities for collaboration. District and school leaders should create and systematically support opportunities for teachers and administrators to collaborate within and across districts and schools, with support from relevant experts, with a focus on how to improve instruction to support students' learning as described in the *Framework* and the Next Generation Science Standards.

RECOMMENDATION 16 Identify, participate in, and build networks. Science education leaders should identify, participate in, and help build national, regional, or local networks that will enable communities of practitioners, policy makers, science experts, and education researchers to collaboratively solve problems and learn from others' implementation efforts. Teachers and administrators should be encouraged to participate in such networks as appropriate.

RECOMMENDATION 17 Cultivate partnerships. Science education leaders should identify partners in their region and community that have the expertise, motivation, or resources to be supportive of their efforts to implement the Next Generation Science Standards and develop relationships with them. In collaboration with potential partners, leaders should determine the kind of support each partner is most suited to provide and develop strategies for working with them.

THE POWER OF COLLABORATION

Collaboration, partnerships, and networks can be powerful mechanisms for supporting the changes called for by *A Framework for K-12 Science Education: Practices, Crosscutting Concepts, and Core Ideas* (National Research Council, 2012; hereafter referred to as "the *Framework*") and the Next Generation Science Standards (NGSS) through sharing expertise and strategies. Networks can include people working within school systems, such as collaboration of leaders across states, or among teachers across schools and districts, or even within schools across grades. Networks also can connect the school or district with external partners.

The NGSS have already been adopted by several states and districts, while others have adopted new standards that share some part of the vision and will require similar changes to instruction, teachers learning, and curriculum. Some states or districts are providing professional learning experiences based on *A Framework for K-12 Science Education* (National Research Council, 2012) while awaiting regularly scheduled standards revisions to consider adopting new standards.

To enhance the capacity of district and state leaders charged with implementing the NGSS (or standards that closely resemble the NGSS) to rapidly share data about effective strategies, materials developed, and results achieved, cross-state efforts will be critical. Such work has already begun with the efforts of Building Capacity for State Science Education, a network organized by the Council of State Science Supervisors, and Achieve's NGSS Network for states that have adopted the NGSS. Such networks have the potential to (1) build and sustain a community of practice among the people who are implementing the NGSS; (2) codify, organize, and share knowledge about effective approaches and practices; (3) serve as a forum for new science education leaders to connect and learn from those with more experience; and (4) provide a locus for scientists and education researchers to connect with science education leaders, both to aid the flow of research-based approaches to the field and to provide researchers a window on problems of practice and of large-scale implementation of demanding new standards that need further study.

Networks among similar schools and districts within a state or across states can be especially helpful for supporting implementation. For example, schools that serve similar student populations or districts that face common challenges of distance or limited resources would be helped by opportunities to share strategies. Similarly, schools and districts that follow similar instructional models or adopt the same curriculum materials can collaborate productively on developing and sharing materials and strategies. Technology can play an important role in facilitating communication and sharing of materials among network members.

At a more local level, district and school leaders can identify other schools, districts, or science-rich organizations in the region that are working on implementation of the NGSS and form collaborations to share ideas and resources. For example, collaboration can allow for pooling of resources across districts to provide special programs focused on supporting novice teachers as they work to implement the NGSS (Weiss and Pasley, 2006). Once partnerships and networks are established, resources need to be devoted to maintaining them, including people to facilitate the collaborations and to help maintain communication among partners. It might also need to include funding to cover the costs of the technology used to facilitate sharing of information and materials (costs of computers, virtual meetings, maintaining a website and file space, etc.). These resources may come from external partners as well as from school districts. Collaborations, networks, and partnerships should be monitored to determine how well they are functioning and make changes when necessary.

NETWORKS FOR TEACHERS

Networks of teachers working together to understand and implement changes in their instruction can be powerful mechanisms for supporting implementation of new science standards (Coburn et al., 2012; Penuel and Riel, 2007). Such networks provide a mechanism for teachers to share ideas about teaching, learning, and assessment; stories about students' successes and difficulties; strategies for managing learning groups; and tips for using technology (Penuel and Riel, 2007).

There are some key features of networks that have been shown to be more effective than others in supporting sustained change in instruction. Effective networks include strong ties (frequent interaction and social closeness), access to expertise, and deeper interactions (focused on underlying pedagogical principles, the nature of the discipline, or how students learn) (Coburn et al., 2012). District policy can shape how teachers engage in networks and whether their participation supports changes in their instruction (Coburn et al., 2013). Policies can support more frequent and deeper interactions and help teachers identify local experts, but they can also disrupt ties, interrupt the flow of resources, and eliminate supports that encourage interaction (Coburn et al., 2013): see Box 7-1.

A study of 21 schools in California engaged in school-wide reforms suggests several additional characteristics of effective teacher networks (Penuel and

BOX 7-1

TEACHERS' NETWORKS AND INSTRUCTIONAL REFORM

Coburn and colleagues studied the role of teachers' networks in supporting instructional reform in a school district adopting a new mathematics curriculum (Coburn et al., 2012, 2013). The district was a midsize urban district that adopted an innovative curriculum in the 2003-2004 school year—the first year of the study. The district also began an initiative to support teachers in learning the new curriculum, including creating school-based instructional coaches and multiple opportunities for teachers to meet with others to talk about mathematics.

The researchers focused on four schools that differed in terms of the strength of the existing professional community in the school and the level of teachers' expertise. All four schools had 70 percent or more of their students enrolled in free and reduced-price lunch programs at the start of the study, and 70 percent or more of their students were Latino, mostly of Mexican origin. About half of the students of all four schools were classified as English-language learners. The researchers focused on three teachers in each school (only two teachers in one school) and collected detailed information about the teachers' social networks.

In the first year of the study, the district adopted a new mathematics curriculum and designed a series of activities to help teachers implement it. First, the district created the role of a school-based mathematics coach, and each school was to appoint a minimum of two half-time coaches/teachers to work with teachers. Second, the district instituted weekly grade-level meetings to facilitate joint planning and biweekly school-based professional development. A district-level team supported coaches, providing them with regular professional development and observing their work once a month. Third, the district provided professional development to select teachers in the summer and intersession.

Riel, 2007). First, getting help from outside of one's immediate circle is valuable for obtaining new information and expertise. Second, making it clear who has expertise to help with a specific challenge is helpful. To do so, it is important to provide venues where people talk about their teaching, as well as publicly recognizing success and achievement in ways that encourage teachers to seek out their colleagues for help and resources. Third, meetings and committee structures that allow teachers to participate in multiple meetings that cut across different functions in the school allow teachers to get different perspectives on the instructional changes they are striving to make. This approach seems to be more effective than approaches where all of the information flows from leaders and administrators to teachers or those where leaders assume each teacher is developing new approaches to instruction without explicitly thinking about the importance of expertise.

In the second year, the district offered additional professional development to teachers in cross-grade settings. The school-level professional development program, instituted the previous year, was changed to cross-grade-level configurations. Professional development also became more focused on how students learn, the nature of mathematics, and how to solve mathematics problems. The district also continued to provide professional development to the onsite coaches, increasing their mathematics expertise.

In the third year, a new superintendent changed the district's priorities and withdrew support for much of the network program. The new superintendent cut time for mathematics instruction, gave schools authority for budget and staffing decisions, and ended the district stipend for mathematics coaches. In response, principals in three of the four schools cut back to a single half-time coach.

During the first 2 years of the program, most of the teachers' networks increased in size and diversity (with more ties to teachers in different grades or different schools). The structures put in place by the district also allowed teachers to learn where expertise was located in their schools and be more strategic in making decisions about who to ask for advice. District policy also influenced the resources that teachers accessed through their networks by providing information and materials that teachers acquired from their colleagues and by providing professional development that increased the level and breadth of available expertise. The coaches also modeled and encouraged ways of talking about mathematics that encouraged deeper interactions among teachers. They talked more about student learning and the details of instruction rather than exchanging quick stories or sharing materials or activities.

In year 3, when the formal supports were withdrawn by the district, the quality of the networks that the teachers developed in the first 2 years of the initiative influenced their ability to sustain the new instructional approaches (Coburn et al., 2012). Networks with combinations of strong ties, deeper interactions, and high expertise helped teachers continue to adjust their instruction even when formal supports were removed.

Finally, freeing up the time of experts to help others is important. Such experts might already be in formal roles that allow them to share their expertise, but they also may be informal leaders who have little time outside of their teaching responsibilities to serve as resources to their peers. Recognizing these informal leaders and giving them time to work with peers can be helpful in building effective teacher networks.

COMMUNITY PARTNERS

A number of resources outside schools are already helping to support science learning (National Research Council, 2009; Traphagen and Traill, 2014). Such resources include museums, science centers, zoos, aquariums, and planetariums.

A survey conducted by the Institute of Museum Services (2002) of approximately 11,000 museums and science centers of all sizes in the United States found that cumulatively, these institutions spent over $1 billion annually on programs for kindergarten through high school (K-12) in 2000-2001 and provided millions of instructional hours for teachers.

A great deal of science learning occurs in out-of-school settings and can complement the learning that occurs in school (National Research Council, 2009). It can be very helpful for state and school district planners to identify such potential partners early in the implementation process and to invite their participation in planning to take maximum advantage of their expertise and facilities and to develop a shared understanding of the vision for science education. The *Framework* and the NGSS provide a common language that can help educators from the formal and informal learning worlds to coordinate their efforts and develop effective partnerships.

Another major resource for supporting student engagement in science and engineering is the vast number of afterschool and summer programs that offer learning opportunities for science, technology, engineering, and mathematics (STEM) (National Research Council, 2009). Afterschool and summer programs and the growing number of science- and engineering-related competitions provide many opportunities to engage children and youth in STEM. Organizations, such as the National Afterschool Association, the Afterschool Alliance, and the Summer Learning Association provide curricula and professional development for leaders of local afterschool and summer programs that are focused on science and engineering. Similar services are provided by major youth organizations, such as 4-H and Girl Scouts. Engagement with these organizations at the district, regional, and state levels can provide opportunities for partnerships that can increase the time that students spend engaging in the practices of science and engineering. Such engagement can also increase the quality and availability of out-of-school experiences that align to the vision of the *Framework* and the NGSS.

Higher education institutions and businesses also can serve as productive partners. Several states already have STEM coalitions, which bring together many organizations and businesses interested in STEM education, which can be tapped. Particularly for science and engineering, more than for reading or mathematics, it may be necessary to look to external partners to bring deep expertise to districts.

STEM professionals from higher education, science museums and centers, and science- and engineering-related industries may already have roles in working with teachers to enhance their understanding of science content. In the context of the *Framework* and the NGSS they can be invaluable resources for helping teachers understand and engage in the science and engineering practices. Both 2-year and 4-year colleges and universities are important potential partners. They often provide professional development opportunities for teachers, and they prepare future teachers. Collaboration is important to ensure that the learning experiences provided for both preservice and inservice teachers are consistent with the vision of the *Framework* and the NGSS (see Chapter 8 for additional discussion of teacher preparation).

Partners vary in the kinds of expertise and resources that they can bring to collaboration. Districts need to be strategic about defining what external expertise would be useful to them and finding partnerships that can provide it. The range of valuable technical expertise and resources is wide: it includes science content knowledge, experience in doing science or engineering work, special facilities or materials, experience providing professional development for teachers, and expertise in research and evaluation. Some partners may provide time and space for students' out-of-school experiences but may need other experts to help make that time effective for the type of science learning envisaged by the *Framework* and the NGSS. Yet other partners can provide monetary or public relations and advocacy support or support continuity of effort. Partners who wish to financially support schools and districts can invest in building and maintaining networks that develop and provide tools and resources for teachers and leaders of science. Science-rich organizations can help provide professional development opportunities for teachers and support teachers in developing and sharing research-based NGSS-aligned curriculum units and curriculum resources.

In the Local Systemic Change initiative funded by the National Science Foundation (see Box 4-1, in Chapter 4), major partners who were brought in as stakeholders included universities, research institutes, and businesses. The most effective external partners provided resources and help in establishing stable structures for sustaining improvements in science and mathematics education, such as centers for disseminating materials and professional development (Weiss and Pasley, 2006).

States and districts are most likely to need to work with external partners in documenting and evaluating the implementation process. The *Framework* stresses the importance of understanding the impact of the NGSS, starting with the first steps in implementation (National Research Council, 2012, Ch. 13). This understanding requires documentation of the conditions for effective implementation as well as documentation of outcomes. The kind of systematic evaluation needed will likely require collaboration with individuals and organizations (including universities) that have the expertise to carry out such evaluations.

BUILDING AND MAINTAINING PARTNERSHIPS

Building successful partnerships requires careful consideration of differences in the priorities that different partners might bring to the table (Firestone and Fisler, 2002; Goodlad and Sirotnik, 1988; Heckman, 1988; Kornfield and Leyden, 2001; Vozzo and Bober, 2001), differences in the status and authority of participants (Bickel and Hattrup, 1995; Coburn et al., 2008; Freedman and Salmon, 2001; Goodlad and Sirotnik, 1988; Osajima, 1989), and clarity of roles (Freedman and Salmon, 2001; Goldring and Sims, 2005; Handler and Ravid, 2001; Hasslen et al., 2001).

It is important to openly discuss and establish clear authority relations and develop a shared understanding of appropriate roles and relationships in order to avoid power struggles and misunderstandings (Coburn et al., 2008). The process of establishing clear roles and authority may require more attention when partnerships are formed at the central district rather than at the school because districts often have multilevel management structures, loose connections between these levels, and high staff turnover (Coburn et al., 2008). Partnerships can begin as informal relationships, but for partnerships to last it is important to formalize them with some organizational agreements and structures and some ongoing activities or periodic meetings.

Potential partners interested in collaborating with schools and districts need to become familiar with the vision and language of the *Framework* and the NGSS. A shared vision, and the shared language around the vision provided by the *Framework*, will strengthen collaborative efforts. Such efforts can then improve the quality of both in-school and out-of-school learning experiences for both students and teachers and make the connections between them more visible. One example of a multipartner collaboration built around implementation of the NGSS is the California K-8 NGSS Early Implementation Initiative, which involves eight school districts, two charter management organizations, and the K-12 Alliance of WestEd[1] (a nonprofit research and development agency working in education),

[1]WestEd is a nonprofit research and development agency working in education, based in San Francisco: see http://www.wested.org/about-us/ [November 2014].

and it was designed with input from the State Board of Education, the California Department of Education, and Achieve.[2]

Partnerships generally require a long-term commitment that must itself be supported by an ongoing financial commitment, possibly from external funders, who may also be partners. External partners may also provide organizational expertise that lends stability to a network and maintains its focus on science learning. External partners can also provide a voice for advocacy, both in the community and for other possible external funders, of the reasons behind the changes in science education and the support and resources needed to maintain those changes.

PITFALLS TO AVOID

Lacking a Common Understanding of the Vision

Successful partnerships and networks need to be guided by a shared understanding of the vision of the *Framework* and the NGSS. Differing interpretations of the goals of reforms can undermine their success (Spillane, 2004). If partners are not working together toward a common vision, then collaborations can be counterproductive. It is important to clarify goals early in the collaboration and continue to revisit the vision.

Having Competing Goals Among Partners

Partners have different roles with respect to education and bring different strengths and expertise to any collaboration. They also may have different goals for their organizations. In order to support implementation of the NGSS, the educational goals of the district or school need to be recognized and accepted by all partners, and the district or school needs to understand the goals of the external partners and ensure that they are sufficiently aligned for the partnership to function effectively to meet the needs of all partners.

Failing to Clarify Responsibilities and Monitor Partnerships

Whether or not funding is involved, any joint undertaking requires that all partners have a clear understanding of each other's roles and responsibility. Without

[2]Achieve is a nonprofit education reform organization that works with states to raise academic standards and graduation requirements, improve assessments, and strengthen accountability: see http://www.achieve.org/about-us [November 2014].

a written agreement, such as a memorandum of understanding, there is risk that the partners will fail to meet each other's expectations. It is important to recognize that there may be power and prestige differences between some potential partner organizations, such as prestigious universities or corporations, and schools and districts. As partnerships develop, it is important to work toward mutually beneficial arrangements with reciprocal benefits for all the actors involved (Goodlad, 1988; Linn et al., 1999; Radinsky et al., 2001).

Together, all partners need to make periodic assessments of how well the partnership is working for all of them. If it becomes clear that the partnership is not working well for one or more partners, all need to discuss what adjustments need to made (including to end the partnership, if necessary).

Failing to Establish Mutually Respectful Relationships and Roles

True partnerships require that all partners respect the expertise and the perspectives and concerns of the others. Partnerships where science experts or business leaders fail to recognize the expertise of teachers and other instructional leaders, or where educators fail to respect the perspectives of business leaders around developing skills for employability, are unlikely to last long. The organizational work to understand and develop a shared respect for the perspectives and expertise of the different partners may appear to be unimportant when more immediate tasks and pressing needs are in the foreground. However, unless it is given attention, differences in perspectives or the lack of mutual respect will limit the effectiveness of any partnership and may even cause it to fall apart.

8

POLICY AND COMMUNICATION

RECOMMENDATION 18 Ensure existing state and local policies are consistent with the goals for implementing the Next Generation Science Standards. State boards or commissions with the appropriate authority should review and revise where necessary state-level policies with regard to teacher certification, graduation requirements, and admissions requirements for higher education to ensure they do not create barriers to effective implementation. District leaders should ensure local policies such as teacher assignment support implementation of the Next Generation Science Standards.

RECOMMENDATION 19 Create realistic timelines and monitor progress. State, district, and school leaders should ensure that timelines for implementing the Next Generation Science Standards are realistic and are clearly understood at all levels of the system. They should monitor the implementation and make adjustments when necessary.

RECOMMENDATION 20 Use *A Framework for K-12 Science Education: Practices, Crosscutting Concepts, and Core Ideas* and the Next Generation Science Standards to drive teacher preparation. Provosts, deans, department heads, and faculty in higher education institutions should review and revise programs and requirements for teacher preservice training and introductory undergraduate science courses to ensure these are responsive to teachers' needs under the Next Generation Science Standards, at both the elementary and secondary levels.

RECOMMENDATION 21 Communicate with local stakeholders. State, district, and school leaders should develop a comprehensive strategy for communicating with parents and community members about the Next Generation Science Standards and the changes that will take place to implement them, including a multiyear timeline, possible changes in students' assessment results, and how science classrooms may be different. The communication strategy should include opportunities for public dialogues in which parents and others in the community can provide feedback and express concerns.

THE ROLE OF POLICY

State education leaders need to be aware of the interplay of various policy decisions, including some not directly connected to science education. Those policies include budget allocation, human capital, accountability, school configuration, use of classroom time, and course requirements. Leaders need to review them to ensure that they are consistent with the goals of *A Framework for K-12 Science Education: Practices, Crosscutting Concepts, and Core Ideas* (National Research Council, 2012; hereafter referred to as the *Framework*) and the Next Generation Science Standards (NGSS) and to change them when they are not.

ADEQUATE TIME FOR LEARNING

Ensuring that students at all grade levels have adequate time and opportunity for science learning needs to be a priority for successful implementation. Because this time competes with priorities and other demands for student time, science learning opportunities have to be a topic of district- and school-level policy and decision making.

The one obvious time issue is the number of years of science required for high school graduation. Given the standards for grades 9-12 under the NGSS, it is difficult to envision any course sequence that offers all students an opportunity to explore these ideas and reach the expected level of competence with them in less than a 3-year sequence. Indeed, for many students it will require 4 years of high school science or opportunities for students to learn some of the science ideas and practices in the context of career and technical education courses. An example that maps the 9-12 grade standards over each year of high school results in three demanding courses with high expectations for student learning (see NGSS Lead States, 2013, App. K).

States or districts that currently require less than 3 years of science for high school graduation will need to consider how this policy affects students' opportunity to learn the science and take steps to ensure that all students indeed have this opportunity. Decisions about course requirements and sequencing, and about the timing and content of the science testing required by the No Child Left Behind Act,[1] have implications for available course options. Course options, classroom space, and resources devoted to science classes all affect the ability of students to effectively engage in the science practices throughout high school. The NGSS represent the minimum standards to be achieved by all students. Options for interested students to take more advanced science courses at the high school level (advanced placement, international baccalaureate, or honors courses) also need to be supported, and such options need to be equitably available to all students. Policies that tightly restrict the number of courses a student may enroll in each semester may limit options for science for ambitious students.

Other state and or local policies particularly affect time for learning in the lower grades, such as policies on instructional minutes for specific subjects or on how English-language learners are supported to learn English. If state or district policies in these areas result in less class time for science, it may mean that some students are not being given the needed opportunity to learn science. Some rethinking of how best to serve the students with particular needs may be necessary.

---

[1]For a discussion of the act and its implications for science education see *Systems for State Science Assessment* (National Research Council, 2006b).

Teachers need to be supported and given the flexibility to work across subject areas to implement instruction that supports student learning across multiple learning goals: that is, science learning and literacy or language learning need to work together and not be regarded as competing options (Lee et al, 2013; NGSS Lead States, 2013, App. D; Quinn et al., 2013). Understanding such shared learning goals is the kind of activity for which school leaders and teachers can learn from others who are further down the path of implementation, through networks and other kinds of collaboration.

HIGHER EDUCATION ADMISSIONS

Higher education systems will need to review and may need to update their admissions policies with regard to science courses, such as what courses are accepted as "laboratory based" (for a discussion of laboratory experiences in high school, see National Research Council, 2006a). Rigorous new courses intended to meet the NGSS must not be excluded from acceptance because the science investigations that they include are chiefly field work, for example, in Earth or environmental science. Engineering design work, as well as science investigations, need to qualify as meeting requirements for practical or laboratory work. The system for review and qualification of new high school courses as acceptable under admissions standards needs to be consistent with the view of science practices elaborated in the *Framework*.

TEACHER PREPARATION AND CERTIFICATION

Science teacher preparation programs (including both traditional programs and alternative pathways into teaching) and certification requirements will also need to be adjusted to better prepare teachers for supporting the NGSS. Secondary teachers will need to be prepared to make stronger connections across disciplines than is usual in the current high school curriculum: for example, expecting chemistry teachers to cover some applications of chemistry in an Earth science context or physics teachers to engage their students in engineering design. In addition, if new course patterns emerge at middle and high school, there may be a need for new credential options for teachers at these levels that are better matched to the courses.

Ensuring that preservice education and alternative certification programs produce science teachers for all levels who enter the classroom prepared to meet the new demands of science instruction will likely require redesigning both science and science teaching methods courses. It may also entail some changes to certification requirements, particularly those used for alternate paths to subject-area certification. Teacher preparation should include participation in the full range of science practices and applying crosscutting concepts. Finally, the assessments required for teacher licensure and the course work needed for subject-area certification need to reflect the types of learning and assessment tasks that teachers will be expected to develop for students. It is important to ensure that the value of learning activities and assessments that integrate the three dimensions of the *Framework* for students also apply to the evaluations used to assess teachers' ability to teach science.

COMMUNICATION

In making any significant change, such as implementing a new approach to science, state and district leaders need to be aware not only that the change will take time to occur but also of the need to communicate effectively with multiple audiences. Districts need to ensure that parents, teachers, and community leaders understand the goals of the new standards, the evidence for why the new approaches are an improvement, and that some struggles along the way are to be expected.

Information sessions for the public should include opportunities for dialogue in which parents and others can ask questions and express concerns: such dialogue is an important part of communication. These information sessions should occur early in the process of implementation so that they can lay the groundwork to ensure that initial "bumps in the road" do not become barriers to acceptance and ongoing public support for implementation. Enlisting key allies, for example, from local industry, to speak in support of the changes from their perspective as employers can be very helpful in building community understanding and support.

Teachers, parents, and the broader community need to understand the expected timelines for implementation, as well as the implications for the assessment system. As changes in assessment may take time, districts should identify how they will be monitoring the success of their transition to new standards and share that information with their communities. State and district accountability policies will need to take these issues into account, and the community (teachers, parents, and the general public) will need to be engaged and educated to understand the rationale for the changes.

In addition to general communication, districts will need to be prepared to support teachers to communicate with and respond to parents or others who object to the inclusion in the curriculum of particular topics such as evolution or the human role in global climate change.[2]

PITFALLS TO AVOID

Assuming Existing Policies Are Adequate to Support the NGSS

Without a systematic examination of how policies at the state, district, and school levels support or impede implementation of new standards for science, there are likely to be policy barriers to effective implementation. This examination needs to include policies that affect state higher education. It also needs to include state requirements for certification and teacher preparation programs to ensure that they are well matched to the teaching expectations of the NGSS.

Failing to Communicate with Parents and the Community

Failure to communicate to parents and the community and to enlist their understanding and support of implementation of the NGSS can lead to resistance to the new standards or unrealistic expectations of how fast it will occur. To be able to sustain change over time, it is essential that districts reach out early on to help the community engage with and embrace the vision for change and the inevitable process of continuous improvement or they will be caught being reactive when something does not initially go as smoothly as expected.

Messages should be developed to specifically reach parents, and events should be planned to engage them in dialogue about the change process. If parents and community leaders are engaged at the planning stage and are informed about the reasons for the changes and the expected timelines to implement them, they are more likely to be supportive.

Being Unprepared for Unintended Consequences

Decisions and policies do not always lead to the desired outcomes, and they can have unintended effects as well as those that were intended. Leaders at every level need to monitor the NGSS implementation and be alert for unintended consequences of both existing and new policies and resource allocations. The challenge for leaders is to recognize when a policy simply needs time to achieve the intended

[2]Two helpful resources on these topics are the National Center for Science Education and the National Academy of Sciences: see http://www.ncse.com/ and http://www.nap.edu/openbook.php?record_id=5787&page=R3 [November 2014], respectively.

effect or whether there is a need for modification or revision, or shifts in resource allocations, to counter undesired outcomes.

Assigning Responsibility without Authority or Resources

To achieve any change, the people charged with implementing it need not only the knowledge, skills, and leadership to pull people together for the work but also the positional authority to do the things they need to do. Science implementation leaders need to be given adequate control of the resources and decisions essential to making the NGSS implementation work. Teacher leaders expected to support learning of other teachers need work time and resources to do that work. At each level—state, district, and school—leaders need to ensure that the plan and those charged with implementing it are backed by sufficient authority and resources to do the work that they are being asked to do.

9

CONCLUSION

What happens in science classrooms is shaped by many factors, and all of them are part of the complex system of science education from kindergarten through high school (K-12). To improve science education to reach all students in all classrooms, plans for implementing the new vision of science education need to be designed with that complex system in mind. Different components of the system—instruction, curriculum, assessment, professional learning—all need to be designed around the goals for science learning described in *A Framework for K-12 Science Education: Practices, Crosscutting Concepts, and Core Ideas* (National Research Council, 2012; hereafter referred to as the *Framework*) and the Next Generation Science Standards (NGSS). The vision and the goals also need to drive policies and practices at the state, district, and school levels and across grade levels. Consistent attention to coherence is essential to successful implementation of the NGSS or any other set of standards for education.

To maintain such coherence, leadership is paramount. Leadership is critical both in science and in districts and schools: everyone needs to understand the vision of the NGSS and actively work to support it. Cultivating teacher leaders who have expertise in science and science pedagogy and can help to mentor their colleagues is also important for supporting the necessary classroom-level changes that will result in better learning opportunities for students.

The work of implementation will be challenging and will take time. Plans for implementation should allow sufficient time for teachers and administrators to become familiar with the new standards and for teachers to become adept at new approaches to instruction. Appropriately sequencing and setting priorities for the many steps in implementation will be essential. For example, small changes in instruction to incorporate scientific and engineering practices are likely to be implemented more quickly than major redesign of an entire assessment system.

Collaboration, networks, and partnerships are powerful mechanisms for tackling the challenges of implementation and for sharing successful strategies. Forming alliances with higher education, business, and other community partners can bring expertise and resources that may not be present in the K-12 system. In building these partnerships and in all aspects of planning and implementation, the unique needs for science have to be kept in the center.

Although there are opportunities to learn from efforts to implement standards in other subjects, such as English language arts and mathematics, science is unique in three aspects. First, in science, the emphasis is on generating and interpreting empirical evidence. Second, there is relatively limited time and attention devoted to the subject, especially in elementary school. Third, there is a relative lack of deep expertise in science and science pedagogy among administrators in the education system.

To plan and implement the NGSS, ongoing communication throughout the process will be critical. Numerous stakeholders—including educators, parents, businesses, higher education institutions, and community organizations—have a vested interest in the well-being and success of children and youth. They care deeply about whether students are being well served: their ideas and concerns about science education and how best to move forward with improving it need to inform the implementation process.

The changes catalyzed by the *Framework* and the NGSS that are just beginning to take hold in districts and schools across the United States promise to provide *all* children with exciting and challenging opportunities to learn science. Working to ensure that some children are not left out and that educators have the supports and resources they need to make the vision a reality requires long-term, coordinated investments by everyone who cares about science, science education, and the future of the nation's children and youth.

REFERENCES

Allen, J.P., Pianta, R.C., Gregory, A., Mikami, A.Y., and Lun, J. (2011). An interaction-based approach to enhancing secondary school instruction and student achievement. *Science, 333*, 1034-1037.

Atkin, J.M., and Coffey, J.E. (Eds.). (2003). *Everyday Assessment in the Science Classroom.* Arlington, VA: National Science Teachers Association.

Ball, D.L., and Cohen, D.K. (1996). Reform by the book: What is—or might be—the role of curriculum materials in teacher learning and instructional reform? *Educational Researcher, 25*(9), 6-8.

Ball, D.L., and Cohen, D.K. (1999). Developing practice, developing practitioners: Toward a practice-based theory of professional education. In G. Sykes and L. Darling-Hammond (Eds.), *Teaching as the Learning Profession: Handbook of Policy and Practice* (pp. 3-32). San Francisco, CA: Jossey Bass.

Banilower, E.R., Boyd, S., Pasley, J., and Weiss, I. (2006). *Lessons from a Decade of Mathematics and Science Reform: A Capstone Report for the Local Systemic Change Through Teacher Enhancement Initiative.* Chapel Hill, NC: Horizon Research.

Banilower, E.R., Heck, D., and Weiss, I. (2007). Can professional development make the vision of standards a reality? The impact of the National Science Foundation's local systemic change through teacher enhancement initiative. *Journal of Research in Science Teaching, 44*(3), 375-395.

Berland, L.K., and Hammer, D. (2012). Framing for scientific argumentation. *Journal of Research in Science Teaching, 49*(1), 68-94.

Bickel, W.E., and Hattrup, R.A. (1995). Teachers and researchers in collaboration: Reflections on the process. *American Educational Research Journal, 32*(1), 35-62.

Blank, R.K. (2013). Science instructional time is declining in elementary schools: What are the implications for student achievement and closing the gap? *Science Education, 97*(6), 830-847.

Borko, H. (2004). Professional development and teacher learning: Mapping the terrain. *Educational Researcher, 33*(8), 3-15.

Borko, H., Jacobs, J., Eiteljorg, E., and Pittman, M. (2008). Video as a tool for fostering productive discussions in mathematics professional development. *Teaching and Teacher Education, 24*(2), 417-436.

Bowes, A.S., and Banilower, E.R. (2004). *LSC Classroom Observation Study: An Analysis of Data Collected Between 1997 and 2003.* Chapel Hill, NC: Horizon Research.

Bybee, R.W. (2013). *Invitational Research Symposium on Science Assessment: Measurement Challenges and Opportunities (Summary Report).* Princeton, NJ: Educational Testing Service. Available: http://www.k12center.org/rsc/pdf/bybee.pdf [November 2014].

Coburn, C.E., Bae, S., and Turner, E.O. (2008). Authority, status, and the dynamics of insider-outsider partnerships at the district level. *Peabody Journal of Education, 83*(3), 364-399.

Coburn, C.E., Russell, J.L., Kaufman, J., and Stein, M.K. (2012). Supporting sustainability: Teachers' advice networks and ambitious instructional reform. *American Journal of Education, 119*(1), 137-182.

Coburn, C.E., Mata, W., and Choi L. (2013). The embeddedness of teachers' social networks: Evidence from mathematics reform. *Sociology of Education, 86*(4), 311-342.

Cohen, D.K. (2011). *Teaching and Its Predicaments.* Cambridge, MA: Harvard University Press.

Darling-Hammond, L. (1995). Changing conceptions of teaching and teacher development. *Teacher Education Quarterly, 22*(4), 9-26.

Dorph, R., Shields, P., Tiffany-Morales, J., Hartry, A., and McCaffrey, T. (2011). *High Hopes—Few Opportunities: The Status of Elementary Science Education in California.* Sacramento, CA: The Center for the Future of Teaching and Learning at WestEd.

Driver, R., Newton, P., and Osborne, J.F. (2000). Establishing the norms of scientific argumentation in classrooms. *Science Education, 84*(3), 287-312.

Firestone, W.A., and Fisler, J.L. (2002). Politics, community, and leadership in a school university partnership. *Educational Administration Quarterly, 38*(4), 449-493.

Fogleman, J., Fishman, B., and Krajcik, J.S. (2006). Sustaining innovations through lead teacher learning: A learning sciences perspective on supporting professional development. *Teaching Education, 17*(2), 181-194.

Fortus, D., and Krajcik, J.S. (2012). Curriculum coherence and learning progressions. In B.J. Fraser, C. McRobbie, and K.G. Tobin (Eds.), *International Handbook of Science Education* (2nd ed., pp. 783-798). Dordrecht, the Netherlands: Springer-Verlag.

Franke, M., Carpenter, T., Levi, L., and Fennema, E. (2001). Capturing teachers' generative change: A follow-up study of professional development in mathematics. *American Educational Research Journal, 38*(3), 653-689.

Freedman, R., and Salmon, D. (2001). The dialectic nature of research collaborations: A relational literacy curriculum. In R. Ravid and M.G. Handler (Eds.), *The Many Faces of School-University Collaboration* (pp. 3-10). Englewood, CO: Teacher Ideas Press.

Furtak, E.M., Shavelson, R.J., Shemwell, J.T., and Figueroa, M. (2012). To teach or not to teach through inquiry: Is that the question? In S.M. Carver and J. Shrager (Eds.), *The Journey from Child to Scientist: Integrating Cognitive Development and the Education Sciences* (pp. 227-244). Washington, DC: American Psychological Association.

Garet, M.S., Porter, A.C., Desimone, L., Birman, B.F., and Yoon, K.S. (2001). What makes professional development effective? Results from a national sample of teachers. *American Educational Research Journal, 38*(4), 915-945.

Goldring, E., and Sims, P. (2005). Modeling creative and courageous school leadership through district-community-university partnerships. *Educational Policy, 19*(1), 223-249.

Goodlad, J.I. (1988). School-university partnerships for educational renewal: Rationale and concepts. In K.A. Sirotnik and J.I. Goodlad (Eds.), *School-University Partnerships in Action: Concepts, Cases, and Concerns* (pp. 3-31). New York: Teachers College Press.

Goodlad, J.I., and Sirotnik, K.A. (1988). The future of school-university partnerships. In K.A. Sirotnik and J.I. Goodlad (Eds.), *School-University Partnerships in Action: Concepts, Cases, and Concerns* (pp. 205-225). New York: Teachers College Press.

Handler, M.G., and Ravid, R. (2001). Models of school-university collaboration. In R. Ravid and M.G. Handler (Eds.), *The Many Faces of School-University Collaboration: Characteristics of Successful Partnerships* (pp. 3-10). Englewood, CO: Teacher Ideas Press.

Hasslen, R., Bacharach, N., Rotto, J., and Fribley, J. (2001). Learning connections: Working toward a shared vision. In R. Ravid and M.G. Handler (Eds.), *The Many Faces of School-University Collaboration: Characteristics of Successful Partnerships* (pp. 59-72). Englewood, CO: Teacher Ideas Press.

Heck, D.J., Rosenberg, S.L., and Crawford, R.A. (2006a). *LSC Teacher Questionnaire Study: A Longitudinal Analysis of Data Collected Between 1997 and 2006*. Chapel Hill, NC: Horizon Research.

Heck, D.J., Rosenberg, S.L., and Crawford, R.A. (2006b). *LSC Teacher Questionnaire Study: Indicators of Systemic Change*. Chapel Hill, NC: Horizon Research.

Heckman, P.E. (1988). The southern California partnership: A retrospective analysis. In K.A. Sirotnik and J.I. Goodlad (Eds.), *School-University Partnerships in Action: Concepts, Cases, and Concerns* (pp. 106-123). New York: Teachers College Press.

Heller, J.I., Daehler, K.R., Wong, N., Shinohara, M., and Miratrix, L.W. (2012). Differential effects of three professional development models on teacher knowledge and student achievement in elementary science. *Journal of Research in Science Teaching, 49*(3), 333-362.

Herold, B., and Molnar, M. (2014). Research questions common-core claims by publishers. *EdWeek, 33*(23), 12-13.

Institute of Museum Services. (2002). *True Needs, True Partners: Museum Serving Schools.* Available: http://www.imls.gov/assets/1/AssetManager/TrueNeedsTrue Partners98Highlights.pdf [December 2014].

Kanter, D.E., and Konstantopoulos, S. (2010). The impact of a project-based science curriculum on minority student achievement, attitudes, and careers: The effects of teacher content and pedagogical content knowledge and inquiry-based practices. *Science Education, 94*(5), 855-887.

Kornfield, J., and Leyden, G. (2001). Working together: A successful one-to-one collaboration. In R. Ravid and M.G. Handler (Eds.), *The Many Faces of School-University Collaboration: Characteristics of Successful Partnerships* (pp. 194-206). Englewood, CO: Teacher Ideas Press.

Krajcik, J., McNeill, K.L., and Reiser, B.J. (2008). Learning-goals-driven design model: Developing curriculum materials that align with national standards and incorporate project-based pedagogy. *Science Education, 92*(1), 1-32.

Krajcik, J., Codere, S., Dahsah, C., Bayer, R., and Mun, K. (2014). Planning instruction to meet the intent of the Next Generation Science Standards. *Journal of Science Teacher Education, 25*(2), 157-175.

Lee, O., Deaktor, R., Enders, C., and Lambert, J. (2008). Impact of a multiyear professional development intervention on science achievement of culturally and linguistically diverse elementary students. *Journal of Research in Science Teaching, 45*(6), 726-747.

Lee, O., Quinn, H., and Valdés, G. (2013). Science and language for English-language learners in relation to Next Generation Science Standards and with implications for common core state standards for English language arts and mathematics. *Educational Researcher, 42*(4), 223-233.

Lee, O., Miller, E.C., and Januszyk, R. (2014). Next Generation Science Standards: All standards, all students. *Journal of Science Teacher Education, 25*(2), 223-233.

Lemke, J.L. (1990). *Talking Science: Language, Learning, and Values.* Norwood, NJ: Ablex.

Linn, M.C., Shear, L., Bell, P., and Slotta, J.D. (1999). Organizing principles for science education partnerships. *Educational Technology Research and Development, 47*(2), 61-84.

Martin, A.M., and Handy, B. (2009). Factors affecting the implementation of argument in the elementary science classroom: A longitudinal case study. *Research in Science Education, 39*, 17-38.

Marx, R.W., Blumenfeld, P.C., Krajcik, J.S., and Soloway, E. (1998). New technologies for teacher professional development. *Teaching and Teacher Education, 14*(1), 33-52.

Michaels, S., O'Connor, M.C., and Resnick, L.B. (2008). Deliberative discourse idealized and realized: Accountable talk in the classroom and in civic life. *Studies in Philosophy and Education*, 27(4), 283-297.

Mortimer, E.F., and Scott, P.H. (2003). *Meaning Making in Secondary Science Classrooms.* Buckingham, UK: Open University Press.

National Academy of Engineering and National Research Council. (2014). *STEM Integration in K-12 Education.* Committee on Integrated STEM Education. M. Honey, G. Pearson, and H. Schweingruber (Eds.). National Academy of Engineering and Board on Science Education, Division of Behavioral and Social Sciences and Education. Washington, DC: The National Academies Press.

National Governors Association. (2010). *Realizing the Potential: How Governors Can Lead Effective Implementation of the Common Core State Standards.* Prepared by the National Governors Association's Center for Best Practices. Washington, DC: Author.

National Research Council. (2000). *How People Learn: Brain, Mind, Experience, and School (Expanded Edition).* Committee on Developments in the Science of Learning. J.D. Bransford, A.L. Brown, and R.R. Cocking (Eds.), with additional material from the Committee on Learning Research and Educational Practice. M.S. Donovan, J.D. Bransford, and J.W. Pellegrino (Eds.). Commission on Behavioral and Social Sciences and Education. Washington, DC: National Academy Press.

National Research Council. (2001). *Classroom Assessment and the National Science Education Standards.* Committee on Classroom Assessment and the National Science Education Standards. J.M. Atkin, P. Black, J. Coffey (Eds.). Center for Education, Division of Behavioral and Social Sciences and Education. Washington, DC: National Academy Press.

National Research Council. (2002). *Investigating the Influence of Standards: A Framework for Research in Mathematics, Science and Technology Education.* Committee on Understanding the Influence of Standards in K-12 Science, Mathematics, and Technology Education. I.R. Weiss, M.S. Knapp, K.S. Hollweg, and G. Burrill (Eds.). Division of Behavioral and Social Sciences and Education. Washington, DC: National Academy Press.

National Research Council. (2006a). *America's Lab Report: Investigations in High School Science.* Committee on High School Science Laboratories: Role and Vision. S.R. Singer, M.L. Hilton, and H.A. Schweingruber (Eds.). Board on Science Education, Division of Behavioral and Social Sciences and Education. Washington, DC: The National Academies Press.

National Research Council. (2006b). *Systems for State Science Assessment.* Committee on Test Design for K-12 Science Achievement. M.R. Wilson and M.W. Berthenthal (Eds.). Board on Testing and Assessment, Division of Behavioral and Social Sciences and Education. Washington, DC: The National Academies Press.

National Research Council. (2007). *Taking Science to School: Learning and Teaching Science in Grades K-8*. Committee on Science Learning, Kindergarten Through Eighth Grade. R.A. Duschl, H.A. Schweingruber, and A.W. Shouse (Eds.). Board on Science Education, Division of Behavioral and Social Sciences and Education. Washington, DC: The National Academies Press.

National Research Council. (2009). *Learning Science in Informal Environments: People, Places and Pursuits*. Committee on Learning Science in Informal Environments. P. Bell, B. Lewenstein, A.W. Shouse, and M.A. Feder (Eds.). Board on Science Education, Division of Behavioral and Social Sciences and Education. Washington, DC: The National Academies Press.

National Research Council. (2012). *A Framework for K-12 Science Education: Practices, Crosscutting Concepts, and Core Ideas*. Committee on a Conceptual Framework for New K-12 Science Standards. Board on Science Education, Division of Behavioral and Social Sciences and Education. Washington, DC: The National Academies Press.

National Research Council. (2013). *Monitoring Progress Toward Successful K-12 STEM Education: A Nation Advancing?* Committee on the Evaluation Framework for Successful K-12 STEM Education. Board on Science Education and Board on Testing and Assessment, Division of Behavioral and Social Sciences and Education. Washington, DC: The National Academies Press.

National Research Council. (2014a). *Developing Assessments for the Next Generation Science Standards*. Committee on Developing Assessments for Science Proficiency in K-12. J.W. Pelligrino, M.R. Wilson, J.A. Koenig, and A.S. Beatty (Eds.). Board on Testing and Assessment and Board on Science Education. Division of Behavioral and Social Sciences and Education. Washington, DC: The National Academies Press.

National Research Council. (2014b). *Literacy for Science: Exploring the Intersection of the Next Generation Science Standards and Common Core for ELA Standards*. H. Rhodes and M.A. Feder, Rapporteurs. Steering Committee on Exploring the Overlap Between Literacy in Science and the Practice of Obtaining, Evaluating, and Communicating Information. Board on Science Education, Division of Behavioral and Social Sciences and Education. Washington, DC: The National Academies Press.

NGSS Lead States. (2013). *Next Generation Science Standards: For States, By States*. Washington, DC: The National Academies Press.

Osajima, K.H. (1989). Building school-university partnerships: Subjectivity, power, and the process of change. *The Urban Review, 21*(2), 111-125.

Penuel, W.R., and Riel, M. (2007). The new science of networks and the challenge of school change. *Phi Delta Kappan, 88*(8), 611-615.

Penuel, W.R., and Shepard, L.A. (in press). Classroom assessment and teaching. In D. Gitomer and C. Bell (Eds.), *Handbook of Research on Teaching (Fifth Edition)*. Washington, DC: American Educational Research Association.

Penuel, W.R., McWilliams, H., McAuliffe, C., Benbow, A., Mably, C., and Hayden, M. (2009). Teaching for understanding in Earth science: Comparing impacts on planning and instruction in three professional development designs for middle school science. *Journal of Science Teacher Education, 20*(5), 415-436.

Penuel, W.R., Gallagher, L.P., and Moorthy, S. (2011). Preparing teachers to design sequences of instruction in Earth science: A comparison of three professional development programs. *American Educational Research Journal, 48*(4), 996-1025.

Penuel, W.R., Frank, K.A., Sun, M., Kim, C., and Singleton, C. (2013). The organization as a filter of institutional diffusion. *Teachers College Record, 115*(1), 306-339.

Pruitt, S.L. (2014). The Next Generation Science Standards: The features and challenges. *Journal of Science Teacher Education, 25*(2), 145-156.

Putnam, R.T., and Borko, H. (2000). What do new views of knowledge and thinking have to say about research on teacher learning? *Educational Researcher, 29*(1), 4-15.

Quinn, H., Lee, O., and Valdés, G. (2013*). Language Demands and Opportunities in Relation to Next Generation Science Standards for English-Language Learners: What Teachers Need to Know.* White paper written for Understanding Language. Available: http://ell.stanford.edu/sites/default/files/pdf/academic-papers/03-Quinn%20Lee%20 Valdes%20Language%20and%20Opportunities%20in%20Science%20FINAL.pdf [November 2014].

Radinsky, J., Bouillion, L., Lento, E.M., and Gomez, L.M. (2001). Mutual beneficial partnerships: A curricular design for authenticity. *Journal of Curricular Studies, 33*(4), 405-430.

Ratcliffe, M., Hanley, P. and Osborne, J. (2007). Study 3 changes in classroom practice: Executive summary. In, J. Burden, P. Campbell, A. Hunt, and R. Millar (Eds.), *Twenty First Century Pilot. Evaluation Report* (pp. 12-15). York, UK: University of York Science Education Group.

Reiser, B.J. (2013). *What Professional Development Strategies Are Needed for Successful Implementation of the Next Generation Science Standards?* Paper written for the Invitational Research Symposium on Science Assessment, September 24-25, Educational Testing Service, Washington, DC. Available: http://www.k12center.org/rsc/pdf/reiser.pdf [November 2014].

Roth, K.J., Garnier, H.E., Chen, C., Lemmens, M., Schwille, K., and Wickler, N.I.Z. (2011). Videobased lesson analysis: Effective science PD for teacher and student learning. *Journal of Research in Science Teaching, 48*(2), 117-148.

Sherin, M.G., and Han, S.Y. (2004). Teacher learning in the context of a video club. *Teaching and Teacher Education, 20*(2), 163-183.

Smith, M.S., and O'Day, J.A. (1991). Systemic school reform. In S.H. Fuhrman and B. Malen (Eds.), *The Politics of Curriculum and Testing, 1990 Yearbook of the Politics of Education Association* (pp. 233-267). London and Washington, DC: Falmer Press.

Sneider, C. (2015). *The Go-To Guide for Engineering Curricula: PreK-5, 6-8, 9-12.* Thousand Oaks, CA: Corwin Press.

Southwick, J. (2013). Depth vs. breadth in a new inquiry curriculum. In C. Sneider and B. Wojnowski (Eds.), *Opening the Door to Physics through Formative Assessment* [peer reviewed monograph]. Portland, OR: Portland State University. Available: http://www.nsela.org/images/stories/publications/Monograph%20Cover-Text_final7-7-13.pdf [November 2014].

Spillane, J.P. (2004). *Standards Deviation: How Local Schools Misunderstand Policy.* Cambridge, MA: Harvard University Press.

Spillane, J.P. (2006a). *Distributed Leadership.* Hoboken, NJ: Jossey-Bass.

Spillane, J.P. (2006b). Primary school leadership practice: How the subject matters. *School Leadership and Management, 25*(4), 383-397.

Spillane, J.P., Reiser, B.J., and Reimer, T. (2002). Policy implementation and cognition: Reframing and refocusing implementation research. *Review of Educational Research, 72*(3), 387-431.

Sun, M., Frank, K.A., Penuel, W.R., and Kim, C. (2013a). How external institutions penetrate schools through formal and informal leaders? *Educational Administration Quarterly, 49*(4), 610-644.

Sun, M., Penuel, W.R., Frank, K.A., Gallagher, H.A., and Youngs, P. (2013b). Shaping professional development to promote the diffusion of instructional expertise among teachers. *Educational Evaluation and Policy Analysis, 35*(3), 344-369.

Supovitz, J.A., and Turner, H.M. (2000). The effects of professional development on science teaching practices and classroom culture. *Journal of Research in Science Teaching,* 37(9), 963-980.

Traphagen, K., and Traill, S. (2014). *How Cross-Sector Collaborations Are Advancing STEM Learning.* Working paper. Available: http://www.noycefdn.org/documents/STEM_ECOSYSTEMS_REPORT_140128.pdf [November 2014].

Trygstad, P.J. (2013). *2012 National Survey of Science and Mathematics Education: Status of Elementary School Science.* Chapel Hill, NC: Horizon Research.

van Es, E.A., and Sherin, M.G. (2007). Mathematics teachers' "learning to notice" in the context of a video club. *Teaching and Teacher Education, 24*(2), 244-276.

Vozzo, L., and Bober, B. (2001). A school-university partnership: A commitment to collaboration and professional renewal. In R. Ravid and M.G. Handler (Eds.), *The Many Faces of School-University Collaboration: Characteristics of Successful Partnerships.* Englewood, CO: Teacher Ideas Press.

Weiss, I., and Pasley, J. (2006). *Scaling Up Instructional Improvement Through Teacher Professional Development: Insights from the Local Systemic Change Initiative.* CPRE policy brief. Philadelphia, PA: The Consortium for Policy Research in Education.

Wilson, S. (2013). Professional development for science teachers. *Science, 340*(6130), 310-313.

Windschitl, M., Thompson, J., and Braaten, M. (2008). Beyond the scientific method: Model-based inquiry as a new paradigm of preference for school science investigations. *Science Education, 92*(5), 941-967.

Yoon, K.S., Duncan, T., Lee, S.W.-Y., Scarloss, B., and Shapley, K.L. (2007). *Reviewing the Evidence on How Teacher Professional Development Affects Student Achievement: Issues and Answers Report, REL 2007–No. 033*. Washington, DC: U.S. Department of Education, Institute of Education Sciences, National Center for Education Evaluation and Regional Assistance, Regional Educational Laboratory Southwest.

BIOGRAPHICAL SKETCHES OF COMMITTEE MEMBERS AND STAFF

Helen Quinn (*Chair*) is professor emerita in the Department of Particle Physics and Astrophysics at the SLAC National Accelerator Laboratory at Stanford University and cochair of the university's K12 initiative. She is a theoretical physicist who has had a long-term engagement in education issues at the local, state, and national levels. She is a member of the National Academy of Sciences and the recipient of the Dirac medal from the International Centre for Theoretical Physics and of the Felix Klein medal from the International Commission on Mathematical Instruction. She is a member and former president of the American Physical Society. She served as the chair of the National Research Council's (NRC) Board on Science Education and on the NRC committees authoring the reports *Taking Science to School*, *A Framework for K-12 Science Education*, and *Developing Assessments for the Next Generation Science Standards*. She has a Ph.D. in physics from Stanford University.

Matthew Krehbiel is the science program consultant for the Kansas Department of Education and the primary contact for that state's participation in writing the national Next Generation Science Standards. Mr. Krehbiel is also a member of the implementation team for the Kansas College and Career Ready Standards for Math and English Language Arts and the career and technical education agriculture and science, technology, engineering, and mathematics pathway teams. Previously, he taught a wide variety of science courses and was the science, engineering, technology academy leader at Junction City High School in Kansas. He is a recipient of the Award for Excellence in Conservation and Environmental Education from the Kansas Association for Environmental Education. He serves on the boards of the Kansas State Science and

Engineering Fair, the Kansas Association for Conservation and Environmental Education, and the Kansas Association for Teachers of Science. He has a B.A. in biology and natural sciences and a secondary teacher certification in general science, biology, and physics, both from Bethel College, and an M.S. in curriculum and instruction from Kansas State University.

Michael Lach is the director of STEM Education and Strategic Initiatives at the Center for Elementary Mathematics and Science Education and the Urban Education Institute, both at the University of Chicago. Previously, he led science and mathematics education efforts at the U.S. Department of Education. He was a charter member of Teach For America, teaching high school biology and general science in New Orleans, and then joined the national office of Teach For America as director of program design. He has been honored as one of Radio Shack's Top 100 Technology Teachers and as Illinois physics teacher of the year. As an administrator with the Chicago Public Schools, he led the district's instructional improvement efforts in science and mathematics and became chief officer of teaching and learning overseeing curriculum and instruction in 600+ schools. He is a current member of the National Research Council's Board on Science Education. He has a B.S. in physics from Carleton College, an M.A. in science education from Columbia University and in education leadership from Northeastern Illinois University, and a doctorate in educational leadership from the University of Illinois at Chicago.

Brian J. Reiser is professor of learning sciences in the School of Education and Social Policy at Northwestern University. His research examines how to make the scientific practices of argumentation, explanation, and modeling meaningful and effective for classroom teachers and students. He also co-led the development of IQWST (Investigating and Questioning our World through Science and Technology), a 3-year middle school curriculum that supports students in science practices to develop disciplinary core ideas. Dr. Reiser worked with Achieve on the design of the Next Generation Science Standards (NGSS) and on the tools to help states implement the NGSS. He is currently collaborating with several state initiatives to design and provide professional development for K-12 teachers as they implement the NGSS in their classrooms. Dr. Reiser is a member of the National Research Council's (NRC) Board on Science Education and served on the NRC committees authoring the reports *Taking Science to School*, *A Framework for K-12 Science Education*, and *Developing Assessments for the Next Generation Science Standards*. He has a Ph.D. in cognitive science from Yale University.

Heidi Schweingruber (*Study Director*) is the director of the Board on Science Education (BOSE) at the National Research Council (NRC). In this role, she oversees the BOSE portfolio and collaborates with the board to develop new projects. She has worked on multiple NRC projects on science, technology, engineering, and mathematics education including codirecting the study that resulted in the report *A Framework for K-12 Science Education*. She coauthored two award-winning books for practitioners that translate findings of NRC reports for a broader audience: *Ready, Set, Science! Putting Research to Work in K-8 Science Classrooms* (2008) and *Surrounded by Science* (2010). Prior to joining the NRC, she was a senior research associate at the Institute of Education Sciences in the U.S. Department of Education and the director of research for the Rice University School Mathematics Project, an outreach program in K-12 mathematics education. She holds a Ph.D. in psychology (developmental) and anthropology, and a certificate in culture and cognition from the University of Michigan.

Marshall S. Smith is a visiting scholar at the Carnegie Foundation for the Advancement of Teaching. Formerly, he held several positions at the U.S. Department of Education in the Obama and Clinton administrations, including senior counselor to the secretary, director of international affairs, under secretary, and acting deputy secretary. Recently, he was a visiting scholar at the Harvard Graduate School of Education. He previously served as a director of education programs at the William and Flora Hewlett Foundation, where he funded projects focusing on education technology, California state education policy reform, and college readiness. He is also a former dean of the School of Education at Stanford University. Dr. Smith is a member of the National Research Council's Board on Science Education. He has an A.B. in psychology from Harvard College and an Ed.M. and an Ed.D. in measurement and statistics from Harvard Graduate School of Education.

Cary Sneider is associate research professor at Portland State University in Oregon, where he teaches research methodology in a master of science teaching program. His research interests have focused on helping students unravel misconceptions in science and on new ways to link science centers and schools, and he has directed more than 20 grant projects, mostly involving curriculum development and teacher education. He has also taught science at middle and high schools in California, Maine, Costa Rica, and Micronesia. He is currently a member of the National Assessment Governing Board, which sets policy for the National Assessment of Educational Progress, known as "The Nation's Report

Card," and he cochaired a writing team to develop performance expectations in engineering at Achieve, which is managing the development of Next Generation Science Standards for the states. He served as design lead for technology and engineering on the National Research Council's *A Framework for K-12 Science Education* and was a member of the Board on Science Education and its predecessor the Committee on Science Education K-12. He has a Ph.D. in education from the University of California, Berkeley.

Roberta Tanner is retired from the Thompson School District in Loveland, Colorado, where she taught physics, math, engineering, and other science courses. She brought Advanced Placement physics and integrated physics/trigonometry courses to the district, and she also designed and taught microcomputer projects, an award-winning project-oriented microchip and electrical engineering course. She spent a year as teacher in residence with the Physics Education Research Group at the University of Colorado Boulder, and she also taught introductory physics at the university. She is a recipient of the International Intel Excellence in Teaching Award and the Amgen Award for Science Teaching Excellence. She served 5 years on the National Research Council's (NRC) Teacher Advisory Council and on the NRC committee that authored the report *Developing Assessments for the Next Generation Science Standards*. She is currently a member of the NRC's Board on Science Education. She holds undergraduate degrees in physics and mechanical engineering at Kalamazoo College and Michigan State University. She has a teaching certificate and a master's degree in education from the University of Colorado Boulder.